Residues

Nature, Society, and Culture

Scott Frickel, Series Editor

A sophisticated and wide-ranging sociological literature analyzing nature-society-culture interactions has blossomed in recent decades. This book series provides a platform for showcasing the best of that scholarship: carefully crafted empirical studies of socioenvironmental change and the effects such change has on ecosystems, social institutions, historical processes, and cultural practices.

The series aims for topical and theoretical breadth. Anchored in sociological analyses of the environment, Nature, Society, and Culture is home to studies employing a range of disciplinary and interdisciplinary perspectives and investigating the pressing socioenvironmental questions of our time—from environmental inequality and risk, to the science and politics of climate change and serial disaster, to the environmental causes and consequences of urbanization and war-making, and beyond.

For a list of all the titles in the series, please see the last page of the book.

Residues

*Thinking through
Chemical Environments*

SORAYA BOUDIA,
ANGELA N. H. CREAGER,
SCOTT FRICKEL, EMMANUEL HENRY,
NATHALIE JAS, CARSTEN REINHARDT,
AND JODY A. ROBERTS

Rutgers University Press

New Brunswick, Camden, and Newark, New Jersey, and London

Library of Congress Cataloging-in-Publication Data

Names: Boudia, Soraya, author. | Creager, Angela N. H., author. |
Frickel, Scott, author. | Henry, Emmanuel, author. | Jas, Nathalie, author. |
Reinhardt, Carsten, author. | Roberts, Jody A., 1976– author.
Title: Residues : thinking through chemical environments / Soraya Boudia,
Angela N.H. Creager, Scott Frickel, Emmanuel Henry, Nathalie Jas,
Carsten Reinhardt, Jody A. Roberts.
Description: New Brunswick, NJ : Rutgers University Press, [2022] |
Includes bibliographical references and index.
Identifiers: LCCN 2021010328 | ISBN 9781978818026 (cloth) | ISBN
9781978818019 (paperback) | ISBN 9781978818033 (epub) |
ISBN 9781978818040 (mobi) | ISBN 9781978818057 (pdf)
Subjects: LCSH: Chemical industry—Environmental aspects. | Pollution. |
Chemistry, Technical—Environmental aspects. | Chemistry, Technical—
Social aspects. | Chemicals—Safety measures. | Environmental policy. |
Environmental sociology.
Classification: LCC TD195.C45 B68 2022 | DDC 363.73—dc23
LC record available at https://lccn.loc.gov/2021010328

A British Cataloging-in-Publication record for this book is available
from the British Library.

References to internet websites (URLs) were accurate at the time of writing. Neither
the author nor Rutgers University Press is responsible for URLs that may have
expired or changed since the manuscript was prepared.

♾ The paper used in this publication meets the requirements of the
American National Standard for Information Sciences—
Permanence of Paper for Printed Library Materials, ANSI Z39.48-1992.

www.rutgersuniversitypress.org

Manufactured in the United States of America

Contents

Preface and Acknowledgments

It is admittedly odd to have seven midcareer scholars coauthor a single, short book. Our collective arose out of many conversations over several years, especially as catalyzed at "Carcinogens, Mutagens, Reproductive Toxicants: The Politics of Limit Values and Low Doses in the Twentieth and Twenty-First Centuries," a meeting organized by Soraya Boudia and Nathalie Jas at the University of Strasbourg in late March 2010. Several years of working in tandem and in smaller collaborations made all of us aware of how much we could gain by pooling our knowledge and disciplinary approaches to better understand the growth of the chemical industry, the environmental changes it has set in motion, and the manifold, often ineffectual, efforts to regulate it.

Rather than compile individual case studies into another edited volume, we decided to write collaboratively. We worked together in person in Providence (April 2016), Princeton (January 2017), Étretat in France (June 2017), Philadelphia (January 2018), Berlin (July 2018), and New Orleans (September 2019). In hours of discussion, we found ourselves stepping out of our comfort zones even as we drew on well-developed empirical material from our own research. The interpretation that emerged is truly collective, reflected in our amalgamated voice. None of us could have written this book alone.

We owe a debt of gratitude to the institutions that funded our meetings or otherwise helped in our endeavor: Brown University and the Institute at Brown for Environment and Society, Princeton University, the French National Research Institute for Agriculture, Food and Environment (INRAE), Université Paris-Dauphine,

PSL University, Université de Paris, the Science History Institute (formerly Chemical Heritage Foundation), and the Max Planck Institute for the History of Science. Findings from the study of Ambler, Pennsylvania, presented in chapter 2 were supported by the Office of the Director, National Institutes of Health (award number R25-OD010521-01); funding for some of the research that informs chapter 3 was provided by an NIEHS Superfund Research Program grant (award number P42ES013660). Parts of chapter 1 appeared originally in Soraya Boudia, Angela N. H. Creager, Scott Frickel, Emmanuel Henry, Nathalie Jas, Carsten Reinhardt, and Jody A. Roberts, "Residues: Rethinking Chemical Environment," *Engaging Science & Technology Studies* 4 (2018): 165–189. A few passages in chapter 4 are reprinted from Angela N. H. Creager, "Human Bodies as Chemical Sensors: A History of Biomonitoring," *Studies in History and Philosophy of Science* 70 (2018): 70–81, with permission from Elsevier. We also acknowledge the Hagley Museum and Library, Environmental Working Group, Summit Realty Advisors, GZA Environmental, Inc., and Nicole and Gérard Voide of the Collectif des riverains et des victimes du CMMP for letting us use their images.

This project has benefited from discussions with colleagues and friends. Among others, we thank our terrific interlocutors at the sessions on residues at the 2019 Society for Social Studies of Science (4S) meeting, especially panelists Ulrike Felt, Christelle Gramaglia, Evan Hepler-Smith, Pablo Jaramillo, Justyna Moizard-Lanvin, Ryo Morimoto, Nona Schulte-Römer, Elena Sobrino, and also those in the audience who offered comments, including Colleen Lanier-Christensen, Abena Dove Osseo-Asare, Gabrielle Hecht, Michelle Murphy, and Sebastián Ureta. At Rutgers University Press, executive editor Peter Mickulas showed unwavering support for this project from the beginning and, toward the end, Kristen Joseph blithely met the challenge of copyediting the glitches and inconsistencies of a multiauthor text with skill and patience. Special thanks to Sara Shostak for two rounds of immensely helpful critiques, comments, and suggestions; if the book's organization works, it is in no small part due to her keen intellectual eye and narrative sensibilities. Four

Princeton graduate students—Pallavi Podapati, Gina Surita, Jack Klempay, and Francesca DeRosa—provided timely assistance in formatting, editing, and proofreading the text; at Brown, then-PhD candidate Tom Marlow created the maps that appear in chapter 3; and Evan Hepler-Smith deserves a shout-out for intellectual support and collegiality since this project was in its earliest stages.

It turned out, for our particular experiment in collaboration, there is simply no substitute for working together in person. Consequently, we especially thank our partners and families for their forbearance and support as we traveled to meet, talk, and write. Doing so provided the opportunity both to build the intellectual foundation of this work collaboratively, in real time, and also to share unexpected moments to ground this work in the everyday lives we all experience in this residual world. During one memorable moment in the seaside town of Etrétat, we shared with a local shop owner the reason for our visit and the nature of our work: "Like the PCBs in the bay! There is no more fishing here now. Maybe someday there will be again." In that moment, we felt both the reassurance of our work's ability to resonate and a sudden surge of urgency. Despite the damage of past actions and inactions, another world is still possible.

We finished our endeavor during the midst of the COVID-19 pandemic, highlighting other, though related, vulnerabilities of our globalized systems. Among those are the deadly consequences of social and racial hierarchies and the inadequacies of national infrastructures for health care and environmental protection. The current global crisis may eclipse the apprehension of the long-term damage inflicted on the natural world by our castoffs. Certainly pandemic-era requirements for sanitary practices in food distribution and medical care have increased reliance on single-use containers, bottled water, and disposables of all kinds. At the same time, crises such as these have a way of unearthing the residues of inequality that were always with us. We hope to build on this moment of awareness about social injustices, tracking their environmental sources and costs, before they are buried once again. Residues are here to stay. It is time to own them in our intellectual, social, and political projects.

Acronyms and Abbreviations

ACGIH   American Conference of Governmental Industrial Hygienists

ADEME   Agence de la transition écologique, the French Environment and Energy Management Agency

AMPA   aminomethylphosphonic acid, a metabolite of glyphosate

ANDRA   Agence nationale pour la gestion des déchets radioactifs, the French Nuclear Waste Management Agency

ANSES   Agence nationale de sécurité sanitaire de l'alimentation, de l'environnement et du travail, the French Agency for Food, Environmental, and Occupational Health and Safety

ASN   Autorité de sûreté nucléaire, the French Nuclear Safety Authority

BfR   Bundesinstitut für Risikobewertung, the German Federal Institute for Risk Assessment

BPA   bisphenol A

BSE   bovine spongiform encephalopathy

CDC   Centers for Disease Control and Prevention (U.S.)

CEA   Commissariat à l'énergie atomique et aux énergies alternatives, the French Alternative Energies and Atomic Energy Commission (formerly the Atomic Energy Commission)

CERCLA   Comprehensive Environmental Response, Compensation, and Liability Act (U.S., 1980)

CFCs   chlorofluorocarbons

CFR   Code of Federal Regulations (U.S.)

CKDu   Chronic Kidney Disease of "Unknown" Origin

CMA	Chemical Manufacturers' Association, the major trade group for the chemical industry in the United States (from 1872 to 1978 called the Manufacturing Chemists' Association, after 2000 renamed American Chemistry Council)
CMMP	Comptoir des minéraux et matières premières
DDT	dichloro-diphenyl-trichloroethane
DREAL	Directions régionales de l'environnement, de l'aménagement et du logement, the French Regional Environment, Planning, and Housing Authority
ECHA	European Chemicals Agency (E.U.)
EEC	European Economic Community
EFSA	European Food Safety Authority (E.U.)
EPA	Environmental Protection Agency (U.S.)
E.U.	European Union
EWG	Environmental Working Group (U.S.)
FDA	Food and Drug Administration (U.S.)
GATT	General Agreement on Tariffs and Trade
GCCA	Perry and Marty Granoff Center for the Creative Arts, a building at Brown University in Providence, Rhode Island
IARC	International Agency for Research on Cancer (WHO)
ILSI	International Life Sciences Institute, an industry-funded lobbying group based in Washington, D.C.
ISO	International Organization for Standardization
LEED	Leadership in Energy and Environmental Design, the main green building rating system
MCA	Manufacturing Chemists' Association, former name of the major trade group for the chemical industry in the United States (1872 to 1978, after renamed Chemical Manufacturers' Association)
NGO	nongovernment organization
NHANES	National Health and Nutrition Examination Survey (U.S. CDC)
NHATS	National Human Adipose Tissue Survey (U.S. Public Health Service)

NRC	National Research Council, part of the U.S. National Academies of Science
OECD	Organisation for Economic Co-operation and Development
OEL	occupational exposure limit
OSHA	Occupational Safety and Health Administration (U.S.)
PCBs	polychlorinated biphenyls
PFOS	perfluorooctanesulfonate, a global pollutant used in fabric protectors and stain repellents
PMN	premarket notification, required by EPA for new chemicals to come to market after the passage of TSCA
REACH	Registration, Evaluation, Authorisation and Restriction of Chemicals (E.U.)
SCOEL	Scientific Committee on Occupational Exposure Limits (E.U.)
STEL	short-term exposure limit
STS	science and technology studies
TLV	threshold limit value (term trademarked by ACGIH)
TSCA	Toxic Substances Control Act (U.S., 1976)
U.N.	United Nations
UNEP	United Nations Environment Programme
UST	underground storage tank
WHO	World Health Organization (U.N.)
WWF	World Wildlife Fund

Residues

1

Residue Properties

Chemicals are bureaucratic animals, easier to quantify than to tame. The evidence is as close as your nearest computer, accessible to anyone with an internet connection and a web browser. Just run a search for, say, "global chemical production" or "international chemical management" and see what pops up: reports. Dozens of them, with statistics on trends, values, and product classes spilling forth from nearly every page.

The cataloging has been going on for decades. Every few years, one or another international governmental organization, environmental group, or industry association—such as the United Nations, or the Green Cross, or the International Council of Chemical Trade Associations—issues a report on the global impact of chemical manufacturing, or chemical pollution, or chemical regulation. Although they differ in topical emphasis—with some focusing on chemical production and value, others on chemical waste and disposal, and still others on political efforts to enhance chemical safety and minimize risk—such reports all seem to agree on a few basic points. First, chemicals are fundamental to the world we inhabit, natural and artificial, even though the ones that get the most attention are engineered or refined in laboratories. Second, these industrial chemicals pervade modern societies and markets to such a degree that the chemical industry today functions as a basic driver of the global economy, the "industry's industry."[1] And third, chemicals' impacts on human health and the environment are extensive, complicated, and far from completely understood.

The following synopsis, from a 2004 World Bank report, is typical and encapsulates elements of all three points:

> All living and inanimate matter is made up of chemicals. Virtually every man-made product involves the use of intentionally produced chemicals in some manner and thousands are produced and new ones developed every year in response to demands for new and improved products. Some chemicals are unintentionally produced as by-products in manufacturing, industrial and combustion processes. Once released into the environment, chemicals may undergo short or long-range transport as a result of natural environmental processes, are transformed into other chemicals and may cause local, regional and/or global contamination, exposure of humans and wildlife and, in some cases, toxic effects.[2]

Another common feature of such reports is the attention they give to very big numbers. Here are a few examples from the same report:

> Tens of thousands of chemicals are in commercial use at any time. . . .

> Hundreds of thousands of products, articles and formulations are currently in the marketplace. . . .

> [The global chemical industry] employs more than 10 million people. . . .

> . . . an estimated US\$ 1.5 trillion in sales in 1998.

> . . . initiatives have begun to generate data and [risk] assessments on thousands of high production volume chemicals. . . .[3]

In statistics like these, but also in charts, maps, graphs, and other visual displays, the rhetorical power of aggregation is clear. With chemicals, quantity counts. A recent study, which will almost certainly find its way into a new batch of reports in the coming years, puts total mass-produced plastics ever generated at 8,300 million metric tons and estimates total plastic waste generated at 6,300 million metric tons.[4] Such figures grab us; they impress, but they also paralyze. What does the amount "thousands of millions of metric tons" even mean? What would constitute a rational societal response to consequences, including hazards, of chemicals dispersed in such magnitude? We cannot help but wonder if addressing the negative consequences of commercial chemicals would disable the global economy that undergirds the affordable affluence of industrial life. Awe at very large numbers can easily transmute into ambivalence. Paralysis can perpetuate chemical business as usual.

Taxonomies are another way that these reports bureaucratize chemicals. Sifting through reports will inevitably turn up any number of categorization schemes. It is not uncommon to find lengthy discussions of different industrial sectors, different national economies, different regulatory mechanisms, different exposure routes, different toxicological endpoints, and of course, different chemical classes. Each set of categories works like a filter, using abstraction to find clarity amid chaos.[5] How else are we to make sense of "tens of thousands of chemicals," as the report above puts it, a simplistic shorthand that substitutes for a dizzying kaleidoscope of chemicals, hazards, toxicities, and risks?

A few examples of some of the most familiar contaminants illustrate why it is hard to formulate any easy generalities about industrial chemicals, which are diverse in their origins, uses, and effects. Polychlorinated biphenyls (PCBs) are synthetic but have so pervaded living bodies and ecosystems as to be as ubiquitous as some natural substances. Mercury is a naturally occurring element, but it is also highly toxic and has been employed in commerce for centuries, its biological persistence and economic value reflected in global flows mapped by industrial ecologists.[6] Formaldehyde is a natural compound as well, created in the metabolism of plants, animals, and microbes, but it is also produced and widely used in industry—and it is carcinogenic. Similar, too, is asbestos, a mineral now so omnipresent in twentieth-century-era construction as to be found in hundreds of millions of buildings the world over. These are just four examples among many thousands, each one a bit different from the next. Grouping chemicals together too much can mask differences, too little can confound. Sometimes differences are simply effaced in the name of efficiency. For example, polycyclic aromatic hydrocarbons (PAHs) have different toxicities but are regulated in the United States as if they were all basically the same.[7] Other kinds of differences matter to how production has developed, and so too the attendant politics of regulation. The industry that supplies chemical elements such as nickel, chromium, and lead (the "nonferrous metals") is organized separately from the synthetic chemicals industry, and the interests of these sectors frequently diverge.[8] We often do not even think of commercially produced metals, many of which are quite toxic, when we think of "chemicals." Lastly, the range of relevant differences keeps changing. The frequent emergence of newly identified contaminants (e.g., nanoelements), as well as the newly recognized dangers of older pollutants (e.g., endocrine disruptors), pose another kind of challenge to developing any comprehensive view, much less just keeping track of familiar hazards. In these ways, the heterogeneity of chemicals combined with their ubiquity can render them strangely elusive.

We have been speaking of taxonomy with the assumption that chemical knowledge exists and the challenge is one of organization, not discovery. Yet when one drills down into the taxonomies of industrial chemicals, one finds numerous gaps in data as well as contested information.[9] Knowledge, or at least publicly available knowledge, is patchy and incomplete—even for so-called high production chemicals. In 1997, the Environmental Defense Fund reported that for the top-selling three thousand chemicals in the United States, basic toxicity testing was unavailable for more than 70 percent of them.[10] Both government and industry were so skeptical that they undertook their own assessments and found that the Environmental Defense Fund had *underestimated* the magnitude of the problem: according to the EPA and the

Chemical Manufacturers' Association, more than 90 percent of high-production chemicals lacked publicly available health data.[11] A joint government-industry program to make this information available was launched but never completed. Toxicity data are only one kind of information about chemicals, but they are not exceptional in their absence in the public realm. In many cases, data exist but are not published due to the fact that "confidential business information" is protected from disclosure by regulation. Add to this the reality that scientific knowledge is not static—toxicology, for instance, has changed substantially in the past few decades—and one begins to see the many challenges that face those who seek to master chemicals by organizing knowledge about them.

Statistical aggregations and taxonomic schemes are only two of the common strategies for disciplining an "unruly technology."[12] As strategies, they embody particular interests, goals, and politics. They facilitate seeing the problems that chemicals present in ways that suggest certain types of solutions as preferable to others: management, not prevention; paying more attention to formulating regulation than to implementation or enforcement. Moreover, the focus on the very large (aggregates) and the very abstract (categorical classes) can cause one to lose sight of chemicals as concrete and (re)active. The political problem of chemical contamination is not only a bureaucratic problem of accounting and rational management but also, as anthropologist Mary Douglas might have put it, an ecological problem of matter "out of place": in the end, chemicals are material things—ever-present, often invisible, yet oddly inescapable.[13]

So in this book, we part company with the report writers and tabulators of statistics and taxonomists. We are not seeking to disparage data with this move. Good data are critical to both environmental knowledge and regulation. But the efforts devoted to the collecting, compiling, and critiquing of data often result in a call for more or better data on which to act. We are interested in understanding how this dynamic itself has become part of regulatory inaction. Our analysis will not offer a new plug-in for existing policy frameworks but will, we hope, help us rethink the wisdom of imagined solutions.

In coming to this issue from within science and technology studies (STS), we are responding to the work of many colleagues who have taken on the urgency, complexity, and scale of contemporary environmental problems. Research on environmental health, politics, and regulation has nucleated around several key topic areas: the science and medicine of exposure and chemical safety, the role of technical experts in government decision-making, actions by industry to forestall regulation by magnifying scientific uncertainty, and the mobilization of environmental justice movements that understand chemical risk to be stratified by race and class, to name a few.[14] For each of these issues, the relevant environmental problems implicate chemical manufacturing and pollution. Since the publication of *Controlling Chemicals* by Ronald Brickman, Sheila Jasanoff, and Thomas Ilgen (1985), scholars who have squarely addressed the industry and its manifold environmental consequences have generally focused on a polluted locality (a mining or fenceline community, or a building), a toxic substance (lead, PCBs, or Bisphenol-A), a type of victim (consumer, worker, resident, ecosystem), or a regulatory regime.[15] This literature has been exceedingly valuable in illustrating the scope and recalcitrance of environmental problems related to the chemical industry but reflects the same kind of segmentation that bedevils environmental regulatory systems.

Recent developments in STS, such as the scholarly attention to materiality and infrastructures, suggest promising avenues for reconsidering the production and control of chemicals.[16] We explore how these approaches might be exploited to *theorize from* chemical domains rather than applying theory to chemical cases. We use "domains" rather informally here to mean the institutional as well as geographic settings where chemicals and their social and ecological impacts are studied, debated, planned, protested, and remediated. These might include medical, legal, scientific, and industrial settings as well as places—a watershed, a mine, a city or neighborhood. Our approach is to *think with* the scientific concepts, experiences, and materiality of chemical leftovers and aftereffects.

For this, we center on the notion of *residues* and attach it to some current trends across the social and natural sciences and the humanities.[17] Residues have physical agency but are identified—and just as often overlooked—through social, political, and technical institutions. They are, in other words, sociomaterial objects.[18] Much like Latour's injunction to "follow the actors" and Casper's corresponding advice to "follow the molecule," we accentuate the discovery potential in tracing paths of chemical residues as they travel through time, space, and social institutions.[19] However, we want to do more than simply describe the circuitous routes residues take. We also want to offer substantive explanations of residues' world-altering powers. To do this we need to pay attention to not only where residues travel but also why they end up some places but not others and what they do along the way. Humans typically set these journeys in motion yet cannot fully control the destinations of residues or the damage they may wreak along the way. Chemical residues possess any number of distinctive properties—we will focus on five—that can advance our understanding of why domains such as science, law, markets, and regulation are unable to contain or contend with the residues generated by production and consumption. So to understand residues' complicated effects in the world, we need to study and compare their patterns of behavior. This means not only trying to make them visible but also investigating the economic and regulatory reasons they are so hard to see in the first place.

Things to Think With

Residue is an old idea. It derives from the Latin *residuum*, meaning "something remaining." The meaning has not changed much over the centuries. A contemporary definition describes residue as "a small amount of something that remains after the main part has gone or been taken or used."[20] In law, this something refers to what remains of an estate after taxes, debts, bequests, and other payments are made. And in chemistry, residue refers to "that which remains after a process of combustion, evaporation, digestion, etc.;

a deposit or sediment; a waste or residual product."[21] Consonant with aspects of all these definitions, the residues we follow in this book are at once by-products of extractive and industrial technology, history, and organization and also catalysts escaped from the lab, the landfill, or the mine and urging into existence new biological, chemical, geological, and sociotechnical worlds.

Beyond this step toward chemical inclusivity, we will use the term differently than regulators or environmentalists might, for residues are also markers of political, economic, and cultural choices. This is clearly seen in the guidelines of the World Health Organization and U.N. Food and Agriculture Organization for pesticides and environmental contaminants, for which a pesticide "residue" is allowed to remain on fresh fruits and vegetables at the market and is considered safe for human consumption.[22] We think residue is a concept that can capture all these meanings whose complexities reflect the multifaceted sociomaterial nature of chemicals.

For STS scholars, the idea of residue holds two distinct and novel attractions. One attraction is the promise of a new method. Like biological tracers, we can use residues to chart the chemical dispersions and transformations set into motion by industrialization, often disclosing unanticipated environmental and social costs of living on what is now a "synthetic planet."[23] Residues are transgressive in material and geographic senses as well as legal ones. They disobey boundaries, appear where they shouldn't appear, alter environments, and enter communities and bodies without permission.[24] Following them around, rooting them out, holding them up to the light to study and apprehend their qualities as sociomaterial objects allows a different world to come into view. As we will show, it is a world we cannot see so clearly if we begin our exploration with the economics of chemical production, the legal studies of chemical regulation, or the chemical politics of sick communities. The other attraction that residues hold for us is theoretical. In transposing a key category from its everyday and scientific usage, we are engaging in what Henry Cowles has called "endogenous analysis," peering closely at our chemical subjects for clues as to how to understand them anew.[25]

In crafting an endogenous analysis of residues, we follow environmental sociologist Stephen Bunker in insisting that a social analysis of nature (in his case, the political ecology of mining and other extractive activities) begins with the physical properties and geographies of raw materials themselves.[26] Here is one example: White pine has a molecular structure that allows it to float on water. This property, combined with the geography of North American rivers, helped nineteenth-century timber companies solve the problem in logging of the increasing distance separating sawmills and cities from the remaining stands of mature trees. Because white pine floats, rivers moved the increasingly remote felled trees efficiently to market. In thinking through chemical environments, we adopt a similar materialism—admittedly, more processed than raw—which we will call *residual materialism*. Residual materialism centers analytical attention on chemical residues in studies of the emergence and consequences of industrialization.

Like Bunker, we will start with properties, but as STS scholars, we emphasize that these properties are at once material and social. We will spend most of this book exploring what residues mean through many different empirical cases. Only at the end, in our concluding chapter, will we put the pieces together, so to speak, to illustrate the value of residual materialism for telling bigger stories, including the global history of the modern chemical industry. In this way, the organization of the book mimics our own collective process of inductively "thinking through" chemical domains and cases, starting empirically rather than systematically.

We have identified five properties of residues that can be distinguished analytically, as we will try to make clear through a few examples, but they are also closely related. Politics runs through all these aspects of residues, a consistent point of their interconnection. The utility of these properties, for us, lies in reanimating residues as a plastic heuristic. We will track residues through the course of this book, following them as they move through time, through space, and into our awareness.

First, the term *residue* is marked from the outset by a kind of irreversibility. "Something remaining" implies past action.

Residues are the results or outcomes of some already concluded processes. They are leftovers. Remnants. In this sense, residues cannot escape their history, and they provide clues for reconstructing the chemical past. This is equally true of chemical products that are not human-made and those that are synthetic. Humans come into contact with lead, a natural but highly toxic metal, because it has been extracted for industrial purposes and permanently redistributed across the environment. The element lead already existed (mostly in the form of mineral lead sulfide), but its geographical distribution has been dramatically altered by human intervention. PCBs, by contrast, did not exist before chemists synthesized them.[27] But once synthesized, PCBs—especially the highly chlorinated ones—persist for decades, even centuries. So although they were banned decades ago, PCBs have not gone away, and their fat solubility means that these toxic residues continue to bioaccumulate in food webs. Michel Serres observes that "pollution comes from measurable residues of the work and transformations related to energy, but fundamentally it emanates from our will to appropriate, our desire to conquer and expand the space of our properties."[28] Serres's work reminds us that the irreversibility of residues proceeds not only from the molecular nature of the chemical substances but also from the politics of production and regulation, and more specifically, of decisions to create toxicants in the first place as well as to ignore the leftovers.[29] As irreversible objects, these residues—both material and political—can't simply be undone, undermining the idea that there is a "pre-" era to which we can return, even if politically and economically powerful actors want it so.[30] As Jens Beckert has argued, capitalism tends to restrict attention to future possibilities.[31] Residues remind us that the past cannot be ignored.

Second, residues are material objects, though often neglected or treated as immaterial. Because residues can be thin, faint, negligible, and difficult to see, they are also easy to ignore, which can be politically useful. Residues' there-but-not-there quality belies not only a material existence but, often, their actual volume. For example, the volume of residues generated from mining and primary processing often far exceeds the final product. In this way, even as mountains of often-toxic "tailings" remain invisible to ultimate consumers, they tend to pile up among socioeconomically disadvantaged populations and already degraded ecosystems.[32] And even when physical quantities of residues are minuscule, they have substantive, empirical presence and political consequence. For example, toxicity dose-response curves are not always linear; sometimes a little bit of a toxicant can cause disproportionate harm.[33] Cumulative and synergistic effects can also confound. We interact with residues all the time, and chronic exposure to a little poison every day can add up to a lot of poison over the course of a human lifetime or—when we are dealing with heritable mutations—over generations.[34] The geographic dispersion of residues can also make an important difference because a little bit everywhere adds up to a lot. At the height of the Roman Empire, coinage forged from Iberian smelters produced enough fugitive lead to coat the globe, raising the planet's atmospheric lead to levels that are still measurable today.[35] In each of these examples, our mistake and, sometimes, our political folly lie in believing that a little bit of a chemical residue is the same as none at all.

Third, residues are slippery. They have a way of escaping modern production and regulatory systems and are often hiding in plain sight. Our standard surveillance systems miss residues when we don't look in the right place or if we don't look for the right thing. Sometimes we simply don't look at all. We know about pesticides on crops, but we miss the pesticides that drift over farm workers and their families.[36] We think we know how to deal with the dramatic oil spills that grab headlines and despoil coastlines, but we are stymied by the slow drip, drip, drip of crescive oil disasters in the making.[37] And even when we know what to look for and where to find it, some residues still slip away. Silver nanoparticles used in common consumer products like shampoos and conditioners wash down the shower drain, ending up in water treatment facilities that tend to do a good job of pulling nanomaterials out of wastewater. But then the silver particles are incorporated into the biosolid waste stream, much of which is recycled as organic nutrients onto agricultural fields. From there, rain and wind wash nanosilver off the fields and back into streams, despite our best efforts to curtail its travels.[38] Nanoparticles are not unique in their slick mobility. With every new generation of chemical instrumentation, scientists can measure ever-more-minute amounts of environmental contaminants, which turn up even in areas of the world not inhabited by humans or otherwise assumed to be pristine.[39] Exhibit B is surely microplastics, now known to lurk in fresh seafood and quite possibly in our kitchen salts, honey jars, and beer.[40] To take another current instance, public knowledge of the hazards of "forever" chemicals such as per- and polyfluoroalkyl substances (best known for their role in Teflon coating) arrived after their residues slipped through our landscapes and into our bodies, too late for preventive policy.[41]

Fourth, residues are unruly, behaving and transforming unpredictably. Once at large, they interact with each other and the environment in complicated ways. Sometimes, of course, these interactions helpfully eradicate residues. Many synthetic materials break down and are completely taken up into open ecosystems, contributing to the biomass and nutrient cycles that perpetuate diverse life-forms. But not all residues rejoin the environment as ecologically innocuous participants, and their persistence provides a kind of chemical record of industrialization, innovation, and consumer choice. Some residues that are not harmful outside human bodies are metabolized into toxic substances once inside them. Moreover, residues team up. Chemical emissions interact with each other and with substances that are in the soil, water, and air, producing pollutants we had not planned—or planned for.[42] To take familiar examples, ozone and acid rain are both formed by chemical reactions of different kinds of industrial pollutants with the environment (sunlight in the first case and water in the second). Residual antibiotics from agriculture and aquaculture that enter the environment as waste contribute to the prevalent problem of resistance; it turns out that heavy metals, frequently a cocontaminant of antibiotics, further select for antibiotic resistance.[43] These interactions pose problems for environmental protection laws, which regulate exposures substance by substance, task by task, and medium by medium. This kind of hypersegmentation in the law does not correspond to a three-dimensional world of mixtures and mobility, in which industrial chemicals and organisms interact in air, soil, and water to generate unexpected entities, environments, and risks.

Fifth, residue is primarily valued for the work it creates, less so for the work it accomplishes. In this way, it has an odd sort of negative identity: matter that is not supposed to matter. To call something a residue means that its useful life is effectively over; that it has aged, moved on, or is otherwise to be disregarded. Residues can be costly to remove, however. When they are labeled as waste, residues require further work or energy to clean up, dissolve, and clear away. It is the unwantedness of residue and the negative attention it inspires that can make it incredibly expensive. To take a "crude" example, the oil lost to British Petroleum, LLC, when it gushed uncontrollably from the Deepwater Horizon oil platform in the Gulf of Mexico in 2010 was then valued at about $250 million.[44] The cost to responsible parties and U.S. taxpayers for stopping the leak and cleaning up the spilled oil is estimated to be $14 billion.[45] But when residues are not identified as waste, they can lurk in invisibility. They are there but not there, like the island-sized patches of plastic debris littering the oceans.[46] In this way, the bureaucratic act of labeling chemicals—as wanted/unwanted, important/unimportant—attaches residues to geopolitics. Labeling practices vary from country to country so that what is marked as a residue in one country or at one time may be recognized as a useful chemical product in another, creating discontinuities in international policy. For example, asbestos is now considered a residue in the few countries that have banned it while it continues to be used as a building material in many other countries. Similar disharmonies are seen in national policies governing the use of lead in the manufacture of paint, pipes, and gasoline.[47] Plastic bags are increasingly seen as objects to be banned from cities when they were still considered useful (if not indispensable) just a few years ago. In this way, bureaucratizing chemicals can amount to an uneven vanishing act, serving as blinders and focusing our attention on which environmental leftovers we notice and which we fail to perceive, which we pay to remove and which we leave alone.

Creators of negative value. Unruly in their interactions with nature and slippery in evading social infrastructures of control. Often mistaken as immaterial yet marked by irreversibility. While there are certainly others, we have followed these five properties as useful guides, helping us scour the landscape for clues about where residues lurk and how they behave. As analytic tools, the five properties are broader and more malleable than the physical properties chemists have identified, relating the behavior of chemicals to molecular configurations of atoms and bonds or the familial qualities of chemical classes that toxicologists and pharmacologists have described in terms of "structure-activity relationships."[48] At the same time, the five properties are also more precise and, we believe, better suited to our chosen task than many of the organizing concepts that social scientists have developed for studying materiality. These include concepts from urban studies (urban metabolism and metabolic flows), environmental studies (metabolic rift, Capitalocene), energy studies (the landscape-regime-niche scheme of transition theory), and STS (cyborgs, rhizomes), to name just a few.[49]

Metaphors such as metabolic flows or metabolic rift draw attention to the inherent interactions among ecological and social systems or landscapes but are overly general for the work we need them to do.[50] For example, we cannot adequately explain the patterns of residues' behavior with analytical methods built to analyze circulation without also focusing on the consequences. By following residues, we can bear witness to coproduction and its effects across different institutional and geographic domains, as residues travel from labs to waste pits and back again.[51]

Other key concepts, such as cyborgs or rhizomes, emphasize biology and the ontological hybridity of living beings.[52] For example, people carrying PCBs in their bloodstreams—virtually all of us—are cyborgs in this usage. Yet the cyborg metaphor does not offer a ready way to differentiate among varieties of cyborgs, such as people exposed to PCBs and people exposed to asbestos. Another relevant approach is the history and sociology of ignorance, which examines gaps in knowledge and the social and political interests arrayed against closing them.[53] While "undone science" remains a critical issue for scientists and scholars concerned with environmental chemicals, the problems on the ground are ontological as well as epistemological.[54] Our attention to the sociomaterial properties of residues allows us to follow them, gauge their effects, and analyze the variations in the patterns we find by using a family of complementary scalable concepts.

We think that attention to residues can help fill in the gaps amid valuable work already underway in many sectors of the academy and beyond. A focus on residues can connect studies on local communities to broader patterns of environmental violence and lead us to highlight its nonhuman as well as human sufferers.[55] We see focusing on residues as complementary to ongoing efforts while opening up other, as yet undefined, possibilities for building new kinds of politics within new kinds of science.[56]

Consider how residues refract the Anthropocene, a topic that has taken environmental studies by storm.[57] In a strictly material sense, residues *are* the Anthropocene—incontrovertible evidence of human activity that has become sedimented into the planet's terrestrial record. Residues are the chemical and elemental signals that physical geologists are using to identify a new epoch, mark its origins in time and space, and trace out the stratigraphic consequences of human history for Earth history. In another, more discursive sense, residues subvert a popular narrative about the environmental costs of industrialization, which has been tethered to the combustion of carbon and atmospheric climate change. Burning carbon creates residue, of course. But this framing misses the other side of the carbon coin: Oil and gas are the feedstocks for chemical industry, which synthesizes a massive variety of compounds used in every sector of the economy. Residues help us see the Anthropocene as the combustion *and* synthesis of carbon-containing compounds, as well as the industrial manipulations of dozens of other chemical elements. These material transformations are rearranging not only atmospheric and ocean chemistry but biology as well, altering the course of evolution, including human evolution.[58]

As this rendering of the Anthropocene illustrates, residue offers a way to reckon with ecological change, regulation, and health in terms of its materiality and irreversibility. It enables us to draw on environmental science while recognizing its limitations, as well as to employ our critical perspective as social scientists. In tracking residues, we necessarily confront the present/past nature of chemical contaminants, the unavoidable issues of scale, and the conundrum of voluminous yet inadequate data. We see residues, presented in terms of their sociomaterial properties, as building capacity within STS to catalyze theories generated from chemical domains and so reimagine and remake our environments.

Itinerary

The same qualities that make residues interesting also present challenges for expository writing. Residue is an idea that suggests many meanings at once—all, or virtually all of them, consequential for human societies and natural systems past, present, and future. How do we write a short book about that? The only way we can: we take shortcuts and hope that readers will be content with our approach for now, and others will take enough inspiration to fill out missing details and rework the interpretation. Our route through chemical environments is more surgical than synoptic. Rather than attempt a comprehensive assessment, we have drawn on empirical work we have done, using historical, ethnographical, and sociological methods. The result is a set of detailed stories that are exemplary but not, we think, essential to the broader picture that residues expose. There are myriad twists and turns in the terrains that residues mark. A completely different set of cases could be assembled to illustrate the same points and patterns and hopefully will be, by others, in future work. We admit at the outset our examples are not representative of the industry overall; we use them to prod and provoke, not confirm or cement. We follow asbestos in every chapter but pay less attention to specific synthetic chemicals, which constitute a large proportion of the chemicals sector. Our selection serves to draw attention to the persistent problems with older products and substances, given the greater familiarity most people have with the toxicity of pesticides, DDT, BPA, and other synthetics. The rest of the book is ordered thematically, not chronologically. Moreover, we do not use the properties of residues enumerated above to organize the book, although we will often revisit these key features. Rather, in the chapters that follow, we explore the meanings and materiality of residue by considering three universal processes: legacy, accretion, and apprehension. These three processes—relating to time, space, and understanding—frame our analysis of residues and bring their sociomaterial properties and their environmental and human consequences into clearer view.

Chapter 2 focuses on the path-dependence and persistence of residues as matter out of time.[59] More so than kindred concepts like temporality or history, *legacy* gives emphasis to the residual. Residues are what remain behind and remind us that with chemicals, as with residues of the industrial age more generally, the past is always with us. The prior activities of human societies, whether two thousand years ago or two days ago, generate chemical residues that end up somewhere else, whether in water, in the air, on land, or in bodies. These activities concern not only materials but also regulatory systems, which usually change by addition or modification, dragging administrative organizations and legal precedents interminably forward. Three examples illustrate how legacies emerge and endure, often complicating the present. First, we look at how the community of a former asbestos production plant in Ambler, Pennsylvania, is coming to terms with the economic and environmental consequences of its past. Our second example focuses on legacies that become embedded in administrative law by analyzing the complex and contradictory attempts to control chemicals through environmental regulation. The attempts by the EPA to implement the Toxic Substances Control Act (1976) foundered in part due to the glacial pace of targeting chemicals of concern and developing judicially robust rules. The challenge of environmental waste provides a third and especially important example of how unmanageable the past often proves to be. In all three, we are attentive to the ways in which legacy threatens to become destiny.

Chapter 3, "Accretion," deals with residue as matter out of place.[60] Residues pile up and slip away, leaving trails or accumulating as deposits multiply across the environment. Attention to the accretive properties of residues draws us to processes of accumulation as residues travel and to the transformations that can occur along the way. As residues move, they may change from bedrock to ore to computer components to electronic waste, for example. Or from a synthesized molecule that is not harmful outside human bodies to a metabolite that becomes toxic once inside. Three examples suggest a range of different accretive dynamics and spatial patterns. We begin with a plot of urban land once used as a gas station and then redeveloped into green space as part of an urban college campus. We consider what happens—or doesn't happen—to displaced residues, over time, as property owners and land uses change, as well as how regulations facilitate the dual accumulation of capital and hazardous waste. From local sites and land uses, we move next to global production and the use of asbestos and its accumulation in buildings and construction sites, onto the clothes and into the bodies of construction workers and then, when work for the day ends, into their homes and the bodies of family members. As we chart this course, we see how a known hazard becomes invisible as it moves from domains of industrial production to those of domestic consumption. Our third example in this chapter follows rare earth minerals used in electronics and the massive environmental and health costs their use entails. The extraction, production, trade, and disposal of rare earths is a global activity that leaves toxic traces—especially in the major supplier, China—at every step. Accretion shows us how, literally and figuratively, residue covers ground.

Chapter 4 takes up residue as matter out of reason, using *apprehension* to guide our analysis.[61] Apprehension refers to human capacities to perceive and make sense of residues as part of our physical environment. It also refers to the emotional and moral character of residues as invoking danger. The material reality of residue challenges our systems of reason-making, especially those involving markets, science, and politics. This chapter explores the double meaning of apprehension: understanding and worry. We use three stories to develop this inescapable duality. The first examines how biomonitoring data has prompted scientists, officials, social movements, and the public at large to reckon with the scale and ubiquity of low-level chemical contaminants. Echoing an earlier question about the vast quantities of plastic waste we posed for rhetorical effect, here we pose more substantive questions about the meaning and consequence of contaminant levels when measured in parts per billion or even parts per trillion. Our second example looks at regulatory apprehension, drawing on occupational exposure limits used to regulate exposure to chemicals at the workplace. An ethnography of the E.U. Scientific Committee on Occupational Exposure Limits (SCOEL) in charge of producing those values illustrates the distance between the precise scientific work of experts to develop sophisticated regulatory tools and the often irregular implementation of these tools by others—an example of bureaucratic slippage that is not addressed by SCOEL. The third example centers on the globally used herbicide glyphosate. What strikes us most here is the divergence in usage and apprehension in different places, such as Sri Lanka and Germany. In the final analysis, we situate the apprehension of residues in the material, political, and scientific conditions that have been developing over the past five decades.

Through the journey we make with residues, we describe a situation with often subtle and sometimes dramatic health and environmental consequences. There are also, undoubtedly, countless consequences that scientists, regulators, and human communities have yet to discover. This is a sobering fact of life in the current residue-laden epoch. And it highlights a central theme that runs through this book's substantive chapters: residues permeate beyond the boundaries of political control. As constructed, implemented, and modified over the past half-century, regulation is an integral part of the system from which residues have issued. There is no obvious way out of this situation. Yet understanding the dynamics and configurations that lead to the current situation is a prerequisite for any attempt to develop alternatives.

Echoing Eric Hobsbawm, our book posits the past as an active force, extending into the present and impacting our future not just by invented traditions but by constructed residues.[62] Having traced the afterlives of industrial stuff, located their chemical transpositions, examined how they escape regulatory control, and scrutinized their social and epistemic unruliness, can we now imagine and realize new and hopefully more sustainable chemical environments? In chapter 5, we lay out what we see as the theoretical implications of our broader framework of residual materialism. We offer a brief illustration of how our accompanying toolkit helps us recast the history of modern industrial society and underscore the role of residues as material agents whose power to change the world comes in part from their characteristic social invisibility. As the history of industrialization and its chemical consequences suggest, awareness of our state of globalized contamination often leads to a nostalgia for purity. That seems to us (and others) a faulty response, as well as an impossible ideal.[63]

Our goal is to nudge academic and public discussion of industrial production and regulation beyond the cul-de-sacs of exasperation, complacency, and despair and toward critical engagement with industry, government agencies, and nonprofits about an environmental future in which industrial pollution and consumer waste can be reckoned with, materially and collectively. Considering our residues—theorizing from chemical domains rather than just applying theory to chemical cases—opens alternative paths for thinking through chemical environments.

2

Legacy

In 1929, Swann's Federal Phosphorus Company in Alabama (USA) began producing polychlorinated biphenyls, a class of 209 individual chemicals marketed under the brand name Aroclors. You are likely more familiar with the acronym for these substances, PCBs. And your body is familiar with them too; all of us carry traces of PCBs in our fatty tissues, residues from their incorporation into countless machines and objects, used and discarded.[1] These oily, waxy, hardy substances resist both heat and flame, making them ideal insulators for electrical equipment. During the middle decades of the twentieth century, as the electrical infrastructure for the growing industrial economies was built, PCBs entered landscapes, workplaces, and households in air conditioners, fluorescent light fixtures, refrigerators, television sets, and electrical transformers. Their insolubility in water and resistance to corrosion led PCBs to be used in carbonless copy paper, paints, sealants, ironing board covers, and plastic bottles.[2] In turn, as these objects were discarded, their constituent PCBs began to circulate through landscapes and waterways.

In the United States alone, 1.5 billion pounds of PCBs were manufactured between 1930 until they were banned by the Environmental Protection Agency (EPA) in 1979.[3] By that time, traces of PCBs had become ubiquitous in the environment, particularly in wildlife. Because PCBs are fat-soluble, their residues are taken up and stored in the fatty tissues of animals such as fish and then travel up the food chain, including into the human diet. As it turns out, PCBs are also highly toxic and often degrade very slowly, some having half-lives of hundreds of years.[4] The growing awareness through the 1960s of the extent of hazardous PCB contamination became a major impetus for the introduction of comprehensive chemicals regulation in the United States in the 1970s. It was, in fact, the only chemical specifically named in that statute, the Toxic Substances Control Act, and one of the few substances ever banned by the Environmental Protection Agency.[5] This did not stop the production of PCBs, only shifted its location, which is currently centered in China and Taiwan. But even in the United States, PCBs cannot be relegated to the past. Four decades after they were banned, fish in bodies of water such as the Hudson River in New York remain so contaminated that they are unsafe for human consumption.

The example of PCBs illustrates well the obduracy of many chemicals and their irreversibility as well as the belatedness of knowledge about their dangers. One U.S. EPA administrator referred to PCBs as the "gift that keeps on giving," one that we cannot refuse to accept.[6] This chapter on legacy will focus on the persistence of residues—their tendency as both unruly and slippery creatures to create problems that our regulatory systems are often unable to anticipate or contain.

Among the various terms that refer to how the past shapes the present, *legacy* denotes inheritance, the absence (often death) of the giver, and the involuntariness of the receiver. Legacies are long-term and usually irrevocable. We choose the term with an appreciation for recent scholarship on cultural understandings of temporality. As Pierre Nora has shown, history reins in and reconstructs the past, tethering it to documents, whereas memory associates, projects, and wells up in groups and communities.[7] For Nora and other historians and sociologists of time, understandings of past and future are social, political, secular, even sacred.[8] Legacy also holds a particular currency in regulatory terminology—namely, that "legacy pollutants" are chemicals that remain in the environment long after release and whose harms were not known at the time of use. "Legacy chemicals" also can refer to products whose prior use grants them an accepted status in a system of oversight without new testing or authorization. Sensitive to all these academic and regulatory meanings, we are interested in a materially grounded, and perhaps less self-conscious and less bureaucratized, understanding of the past. By using *legacy*, we emphasize the remains of *longue durée* industrialization, a process that includes social and political decisions (and nondecisions) to acknowledge or ignore environmental contamination. But legacy is more than material (after)life; legacy is lived through the cultural meanings of place and community and the routinization of the social systems—regulations, zoning, and evaluations of health, among others. Legacy is inherited but continuously reinforced.

This chapter uncovers the legacies of residues through three examples: a postindustrial U.S. community reckoning with its environmental remediation; the growth of multilayered, often self-contradicting regulatory regimes for commercial chemicals; and the intractable, global-scale accumulation of waste, from both industrial production and consumer society. The cases move in a long chronological arc from production site to regulation to waste. This ordering of our stories brings out the idea that regulation itself is a reaction to residues, making it difficult for regulatory systems to rectify past damage or contain their global spread. Legacy highlights the irreversibility of our chemically saturated lives and insists that there is no reset button.

Community

Well, I think that there are a lot of people who probably don't know where Ambler is unless you say, "Oh, well, remember, have you ever heard of the White Mountains?" And, you know, a lot of people will say, "Oh, yeah, aren't those those really bad, toxic asbestos piles?" So I think that's one vision that a lot of people have of Ambler. Of course, when you're coming through on the train, that certainly is a lot of what you see. So you add that to the sort of sense of it being almost an abandoned factory town, because from the train, you don't necessarily see that a block and a half away there are all sorts of wonderful restaurants and stuff like that. [. . .] There are a lot of people who think [the asbestos pile is] kind of what Ambler is.

—Beth Pilling[9]

The small community of Ambler, Pennsylvania, is the town that asbestos built. As the quote above suggests, the landscape bears the scars of an industry that ruled and reordered the town and its environs over much of the twentieth century. But the residues left behind have permeated more than buildings, water, and soil. Ambler residents and other stakeholders, like Beth Pilling, the public administrator quoted above, have struggled for decades to remake their town in ways that often bend to asbestos's unruly tendencies. For different people caught up in the protracted struggle, discussions about what Ambler can and should be set the past in contention with the future, the undeniable and irreversible legacies of history against new visions of what the town might one day (again) become.

Keasbey and Mattison Company opened operations in Ambler in 1881 and quickly grew to be one of the world's primary producers of asbestos-containing construction materials. The pipes, shingles, tiles, and bricks produced in vast quantities at the sprawling factory complex were designed with fire-proofing properties to quell the fears of urban dwellers the world over. A chemist by training, company president Dr. Richard V. Mattison used Fordist social engineering to enhance the symbiosis of town and company, including racial-ethnic and class-ordered housing that placed African American and immigrant Italian line workers next to the factory, with managers' larger homes further from the commotion and pollution.[10] As factory production expanded in the early decades of the twentieth century, so did asbestos residues, which began to pile up behind workers' homes along the banks of the adjacent Wissahickon Creek, effectively encircling workers' families (as shown in figure 2.1). As the residues of industrial

FIG. 2.1. Keasbey and Mattison Company complex, showing Ambler's "White Mountains," 1937. Dallin Aerial Survey Company. Hagley Museum and Library. Note the worker housing sandwiched between the waste pile and the factory complex. Courtesy of the Hagley Museum and Library, HagleyID 70_200_09339.

production spilled beyond the factory gates and into the living and recreation spaces of the community itself, what had once been a natural floodplain draining into the local creek grew with time to become two waste repositories: a fifty-foot mound of broken and discarded factory refuse and another raised plane constructed of waste next to the creek. The "white mountains" of Ambler and the adjacent reservoir just north of the hills that also contained a waste site were viewed by many at the time as everyday features of the local environment, including park pavilions where families could gather for picnics in the summer and children could go sledding in the winter.[11]

Yet over the decades, Ambler's economic trajectory would follow a path similar to other twentieth-century industrial communities. The once-robust industry town found itself hobbled by the Great Depression in the 1930s. And though the Second World

War breathed some new life back into the factory, years of decline had already prompted its sale to a British firm, Turner & Newell. Suburbanization following the war led to new markets for asbestos but also increased competition from new chemical products—such as polyvinyl chloride (or PVC, also used in pipes and siding)—which cut into market share. The health case against asbestos was also emerging. By the 1950s, evidence connecting workplace exposures to asbestos and respiratory diseases was mounting and increasingly difficult to ignore. In 1962, the factory closed for good. The impact was immediate and pronounced, shifting residents' cultural understanding of home from one of relative prosperity to one of increasing precarity. As resident Robert Adams recalled of the 1970s, "The plant had closed. Things kind of economically went south for a while from that. There were a lot of boarded up shops on Main Street."[12] Tellingly, those main street shops were replaced by shopping malls built outside the town to serve residents of newer suburban developments.

The passage of the Comprehensive Environmental Response, Compensation, and Liability Act (or CERCLA, and more popularly known as Superfund) in 1980 brought new urgency to the town's historical connections to asbestos and new opportunities to reckon with the legacy of contamination. Superfund made available, in theory if not always in practice, funding to remediate sites considered by the EPA to have significant levels of contamination. Ambler's "white hills" fit the bill. Thus classified, the local meaning of the hills again changed from a sought-after cultural resource to a place of danger and environmental risk. Cleanup of the Ambler Asbestos Piles, as the hills became known, spanned a full decade (1986–1996) and involved "capping" the Piles with fresh topsoil, replanting grass, and encircling the area with a tall chain-link fence with posted warning signs to keep people off the site. This in situ remediation strategy ultimately failed to keep locals from sneaking onto the site and, as importantly, in the view of some failed to curtail erosion of the Piles as concerns of asbestos moving into the Wissahickon Creek watershed emerged.

For many local residents who had endured the first remediation, a second remediation effort was not immediately welcomed. Nonetheless, in 2005, a proposed condominium development on the remaining asbestos-contaminated parcel forced Ambler residents to revisit a past that had been literally buried just a decade earlier. While the condominium proposal was eventually withdrawn, the gears of community and regulatory action had meshed and begun to turn. In 2009, the second Ambler asbestos site—known as the BoRit site—was added to the Superfund's National Priorities List and slated for immediate remediation.[13] Reactions among community members were mixed. Some, like Fred Connor, saw an opportunity to reclaim greenspace for the community: "We think . . . it's important for the township as a whole, the neighborhood of West Ambler, for quality of life, as an economic driver for the homeowners, the property owners there, the residents of West Ambler, not to have a Superfund site in their backyard, but to have an active, attractive park for recreational activities."[14]

Others held more skeptical views. Beth Pilling, quoted earlier in this chapter, cautioned that "governmental employees and scientists have to adhere to . . . a specific protocol that in some ways really does seem like they are not telling you everything, they are not being as forthcoming as they could be, or that they're not as accessible as they could be."[15] Another resident, Salvatore Boccuti, concurred: "My impression of the EPA . . . is that they will cover [the asbestos] up and then test it and find that the test shows that there's no danger, and of course there's no danger. They just covered it up. Once they find there's no danger, they'll go away, and they come back every five years and retest it."[16] Striking a more suspicious tone, Sharon Cooke-Vargas said, "I don't know how asbestos stops at a fence, and [EPA officials] claim that on this side of the fence there's asbestos, but on this two inches on the other side of the fence, there is no asbestos."[17] She also noted that there were no African Americans working on the EPA contracts, questioning whether her community would benefit from the expensive remediation project.[18]

EPA's second cleanup effort in Ambler did more than unearth the material legacy of the town. It also perpetuated a Depression-era narrative that Ambler is a "boom and bust" town. The power and presence of this narrative, scoured from media coverage and interviews covering the period immediately preceding the proposed development project to the time preceding action taken by the EPA, evinces a cultural identity associated with the economic risks of industrialization. But for many, the narrative's cyclical movement of history is disrupted by a more linear, declensionist history invoked by the irreversibility of asbestos contamination. This disjuncture is reflected in an interview with community activist Ruth Weeks who recalled, "I think what [the new remediation plan] did is it brought up the issue to let them know, 'don't try to do that there.' . . . You can't really put anything there, because you can't really clean it up."[19]

Today, visiting the newly renovated Boiler House on the old factory site requires standing, literally, on the residues of the past industry that occupied that space (see figure 2.2). This building, designed originally to create the slurries of asbestos and ceramic that would become the products of this factory complex, is now a LEED-certified platinum office complex. Built into and below the concrete foundations of the building are the remains of asbestos waste that could not be hauled away. But its immediate proximity to the local train station also has meant that the soil there contained the residues of many other prior, coincident, and supporting industries. (Re)developing the space for the twenty-first century meant literally digging through the past and deciding what could or should be done with it. The developer, John Zaharchuk, quickly summarizes in just a handful of sentences what more than a century of industrial use has left behind:

We had high levels of arsenic, lead, and benzo(a)pyrene in the soil. We also had a buried [. . .] rail car that was full of number six oil. So we had to, you know, get the six oil out, remove the tank, properly dispose of it. But it was pretty large. The other soil constituents, not uncommon for a building that's been in

FIG. 2.2. The newly renovated "Boiler House" in Ambler, Pennsylvania, now a LEED-certified platinum office complex. Reproduced with permission of Summit Realty Advisors.

industrial use to have high instance of lead and arsenic. And as far as benzo(a)pyrene, that's actually uncombusted coal, so it dated back to when they had coal here [. . .].[20]

In this way, the seeds of past manufacturing created a negative economy in the present, requiring large sums of capital to remove, transport, or otherwise deal with the toxic remainders of an earlier era's economic production. The shift from a manufacturing economy to a service economy required a careful negotiation of finances as the residues of once-lucrative industries now took on a different kind of value—as something to be cleaned up, at high cost, rather than exploited. Zaharchuk continues, "The cost of transporting three hundred thousand cubic feet of material has got to be entirely cost-prohibitive. [. . .] So from just a financial feasibility standpoint, I think capping it in place, and maybe taking some of the money that you would have spent transporting it away and enhancing the environment, or providing some different amenities, would probably be a better use of funds than wasting a bunch of time transporting material and just taking a problem from this location and moving it to yet another location."[21]

Cast in concrete, these chemical leftovers now form the foundation for a different kind of economy in a different kind of Ambler. The LEED Platinum-certified building, which once housed the boiling operations for the materials embedded into fire-resistant construction materials, counts among its tenants companies specializing in e-commerce platforms, advertising, and law firms—markers of the new industries poised to boom in the twenty-first century as the bust cycle of the twentieth continues to grow in time and scale. The building is both an expression of a shifting economic landscape and a symbol of how the past is remade. As the developers behind the Boiler House reconstruction describe it, "Ambler has risen beyond the stigma of its industrial past to earn recognition as a charming, quaint and welcoming Philadelphia suburb offering great food, shopping, and entertainment."[22]

However, for residents of Ambler, the past is not so easily left behind. For Anne McDonough, the asbestos plant "was the root of the community."[23] But even more, for her, the past is present in the sense that "it has continued to be a focus of who we are as a community, and it has helped people be aware that we are a community, that this is something that impacts us all."[24] Bound up, irrevocably, in Ambler's sense of itself, the legacy of residues has become an unavoidable part of the community's culture. She continues, "It doesn't matter, you're concerned up here at the top of the hill as you are down there, when it's in your backyard. It's a small community, so it does impact us all. We need to be aware of it. We need to make sure that the numbers are not indicating that it's a bigger problem than we're accepting it to be. And again, that's not letting the issue go."[25]

Communities like Ambler struggle to forge new futures for themselves, often repeatedly, from pasts that are intimately bound up with industrial residues.[26] They are consigned to work from inadequate scripts. Even when remediation removes health risks (and we will not try to pronounce on the success of the EPA's efforts in this case), economic activity and community identity rely on reinvention—a cultural manifestation of the legacy of residues.

Regulation

The past not only remains in the residues of our soils, waters, and our bodies, or as cultural scripts. It can also be excavated, fossil-like, from the political and policy infrastructures for governing the production, use, disposal, and reuse of these materials. In general, government regulation of the production of chemicals and the disposal of accompanying waste has *followed* industrialization, not preceded it.[27] The result is often a lag between pollution and prevention and an accompanying disconnect between then and now. Where new regulation is forward-looking, remediation is often understood (mistakenly, in our view) as a problem anchored in the past.

Our brief overview of the expansion of environmental regulation will emphasize how national and local governments and, by the late twentieth century, international bodies contended with the demands and consequences of the growing chemical industry. In doing so, these regulatory bodies were often grappling with the messiness of material contamination—the fact that once residues escape their points of origin, they are exceedingly difficult to contain. But there is another aspect of the residual in the history of chemicals control. The emergence of regulatory action has happened in a gradual, cumulative, often irregular way, in which preexisting agencies and organizations have picked up specific problems related to the management of chemicals at a moment of heightened public concern, often using tools already in existence. An appreciation for the haphazard, fragmented nature of these governing mechanisms is important for helping to dispel notions that this process has created anything like a rational regime of governance. Instead, what remains are the residual effects and intentions of these moments of inspired regulation. Environmental laws and amendments each represent a snapshot of political, material, cultural, and epistemic forces that, once knotted, continue to tether us to that legacy.

The gradual, stepwise emergence of chemicals regulation has led to a system that is highly fractured and multilayered.[28] In the United States, for instance, many different government agencies are involved in the control of chemicals, often with very different aims. The U.S. Department of Agriculture, for example, is charged with promoting the production of crops and other food-stuffs, for which pesticides, herbicides, fungicides, insecticides, and rodenticides are viewed as essential tools. Water quality, to take a different example, was an early target of municipal and state regulation due to the late nineteenth-century threats of typhoid, cholera, and other waterborne bacterial diseases concentrated in cities. The implementation of water purity standards (facilitated by the growing use of chlorine) and the construction of wastewater treatment plants led to the almost complete elimination of those bacterial diseases.[29] However, these efforts did little to address the pollution to both air and water from industrial wastes. In the United States, the federal government did not become a major authority in controlling the pollution of water and air until after World War II, doing so only haltingly until the 1960s.[30] Rachel Carson's *Silent Spring*, published in 1962, became a touchstone for American discontent with government inaction on environmental pollution. Smog, nuclear waste, and pesticides posed real but varied threats not only to natural beauty but also to human health. Fear ran high that industrial pollutants caused a large proportion of human cancer.[31] Public agitation about environmental degradation spurred the updating or passage of seven major pollution control laws in the United States during the 1960s and 1970s.[32]

Richard Nixon consolidated federal oversight over pollution by creating the U.S. Environmental Protection Agency. The founding of the EPA in 1970 is generally cited as a major advance in environmental regulation, but its legal justification is weak, as there is no single piece of legislation setting forth its mission. Instead, the agency oversees the implementation of many unrelated laws passed at different times for different reasons.[33] In this way, the

EPA effectively became an institutional means of managing and perpetuating legacy regulations even as administrators and staff set about trying to define a new era of environmental protection. One early exception was the opportunity for the agency to develop and institute a more comprehensive approach to the regulation of chemicals in commerce. The Clean Water Act and the Clean Air Act set out specific goals for these two aspects of the environment, but existing regulations of terrestrial chemicals remained highly fragmented. Exposures in the workplace had long been managed through private industry standards and, after 1971, also by the U.S. Occupational Safety and Health Administration.[34] Pesticides were regulated under the Federal Insecticide, Fungicide, and Rodenticide Act (administered by different executive agencies over time), and the regulation of food additives and preservatives, as well as drugs, were overseen by the Food and Drug Administration. In 1972, the Consumer Product Safety Commission established safety standards and, when needed, mandated recalls for around fifteen thousand consumer products. Each of these agencies had its own priorities and scientific standards for safety. Chemicals not falling under any of these frameworks were essentially unregulated by the federal government.

The Toxic Substances Control Act (TSCA), passed in 1976, was intended to provide comprehensive oversight of the more than sixty thousand chemicals on the market at that time.[35] It did not rectify the fragmentation, exempting substances regulated by twenty-odd other federal laws (e.g., pesticides or drugs, which are regulated under statutes specifically aimed at those uses).[36] However, it did reflect a new mindset for chemicals regulation. As one industry commentator put it, "the Toxic Substances Control Act is a new way of looking at environmental problems, a systematic and comprehensive approach, not limited to pollutants classified by their occurrence, as in air or water. TSCA contemplates the flow of potentially toxic substances from their origin, through use, to disposal."[37] In the end, however, the bill that was signed into law introduced few (often no) controls over existing chemicals and limited EPA actions on new products.

The bill was introduced in 1971 but not signed into law until 1976, after a chemical accident at a neurotoxic pesticide plant in Virginia once again heightened public pressure for better oversight. Once the Ford administration made it clear that some version of TSCA needed to be signed into law due to this scandal, representatives for the Manufacturing Chemists' Association (MCA), a trade group representing the chemical industry, began hammering out the details of a bill with congressional staffers. James T. O'Reilly, an industry lawyer who actually helped write the provisions, has said, "The 1976 Toxic Substances Control Act (TSCA) contains such obscure and inconsistent phrases that its supporters were doomed to frustration."[38] O'Reilly has referred to TSCA as a "failed statute" and to trying to interpret the bill as "torture."[39]

In general, TSCA was about control of chemicals, not outright bans. There was one exception: banning PCBs was part of the bill. For the EPA, halting PCB use in the United States was the easy step.[40] Figuring out how to deal with PCB contamination already in the ground or the water was (and remains) a huge and intractable problem.[41] For the tens of thousands of chemicals besides PCBs, the statute made the EPA responsible for identifying and controlling toxic substances. However, the agency faced numerous procedural hurdles in fulfilling its mandate to regulate chemicals. These hurdles were not oversights but compromises made to produce a bill acceptable to industry, rendering the law "far less effective than its sponsors hoped."[42] Toxic substances were to be controlled on a substance-by-substance basis, and for EPA to demand that industry supply toxicological information, it was required that the agency provide evidence that current data were inadequate, testing was necessary, and the chemical actually posed an "unreasonable risk." The legal scholar Kevin Gaynor, who analyzed the law shortly after it was enacted, called it "a regulatory morass."[43] Even its provisions "ensuring transparency of safety data" became "rigid procedural handcuffs."[44] This was a statute designed to make industry oversight difficult.

One of the key features of TSCA was the statutory distinction between new and existing chemicals. Legacy chemicals, as

How the Toxic Substances Control Act Evaluates Chemicals
Existing Chemicals

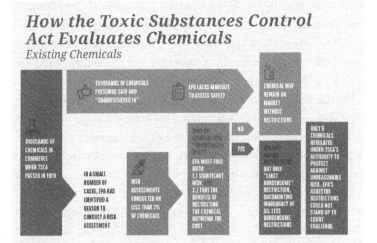

FIG. 2.3. A flow chart illustrating how the 1976 Toxic Substances Control Act regulated chemicals that were already on the market before it became law. *Source:* Environmental Working Group, https://www.ewg.org/key-issues/toxics/chemical-policy.

chemicals with presumably long histories of use and industry safety experience, were deemed innocent until a case for concern could be developed. EPA was reliant upon an exogenous force—an accident or the development of a comprehensive data set generated by a third party, for example—to develop a strong enough case to take any regulatory action, such as requiring toxicity testing (see figure 2.3). This system, in essence, gave companies an incentive to keep producing older, sometimes more environmentally dangerous, chemicals. Despite the chemical industry's rhetoric of innovation, older materials have continued to be produced on a massive scale, in part due to their regulatory status.[45]

For new chemicals, TSCA required companies to file with EPA a premarket notification (PMN), including whatever toxicity test data were available at that time.[46] However, less than 20 percent of PMN's filed included any toxicity information.[47] The EPA had only ninety days after a PMN reached the agency to request additional data (or more time) from the submitting company, and any request for data had to be substantiated with evidence of "unreasonable risk."[48] The agency rarely required additional health or environmental data. Once the ninety days elapsed, that chemical joined the "existing" chemicals inventory and so would not require even a PMN from other producers.[49]

In the end, EPA's oversight reinforced rather than replaced a legacy chemical sector, one built around older products for which regulatory compliance was relatively easy to achieve and in which the regulatory burden for new chemicals was not onerous. Industry had much to gain from the new system. The very existence of TSCA and the EPA's Office of Toxic Substances constituted government action, relieving political pressure on the U.S. government to enact more stringent pollution control measures. Had the EPA been funded to undertake extensive chemical testing itself, or had the courts upheld EPA rule-making, TSCA might have fundamentally challenged the status quo of chemicals production and the associated toxic exposure and environmental contamination.[50] But especially once the EPA adopted a risk assessment and management framework in which economic costs and benefits would be critical to its decision-making, the economic centrality of chemicals made more stringent regulation unlikely.[51] John Warner, the scientist who has helped to define the new field of "green chemistry," estimates that 60 percent of chemicals on the market today are dangerous and have no safer alternative in terms of a substitute; these are chemicals in compliance with existing U.S. law.[52]

Three years after TSCA became law, the Council of the European Economic Community (EEC) updated its regulation for new chemicals, in the Sixth Amendment to the Directive on Dangerous Chemical Substances (79/831/EEC). The Sixth Amendment, like TSCA, exempted many chemical substances, but in other ways, it departed from U.S. law. Its provisions were not organized around "existing" versus "new" chemicals, and it required companies to file a premarket (not premanufacture) notification for their products.[53] Both testing and follow-up reporting were integral to the Sixth Amendment's premarket notification process, although it was left to individual member states to impose specific requirements.

By the 1980s, these differing environmental regulations were viewed as "nontariff" barriers to the growth of the multinational chemical industry. This was of special concern to the Organisation for Economic Co-operation and Development (OECD), which existed to promote free trade across national boundaries, originally just in Europe but eventually worldwide.[54] Early on, the OECD prioritized environmental pollution as a key problem for both economic growth and diplomatic relations.[55] For example, in Scandinavia, acid rain had its origins in British heavy industry, resulting in significant tensions between the countries. The OECD played an important role in the negotiations and resolution of the acid rain problem, as part of the organization's involvement in transboundary chemical pollution.[56]

In March 1971, the OECD formed a "Sector Group on Unintended Occurrence of Chemicals in the Environment," which in 1975 was renamed the Chemicals Group. The Chemicals Group consisted of representatives from the national administrations of the world's most industrialized countries, and they initially organized their work around specific chemicals of concern, such as PCBs and mercury.[57] The Chemicals Group also reviewed existing national legislation on chemicals control, particularly on environmental contamination and premarket regulatory requirements, and began compiling statistics on the sales of chemicals and chemical products.[58]

The OECD was explicitly committed to economic growth and so viewed the issue posed by chemicals as one of how to maintain production and trade while responding to concerns about human welfare and environmental contamination. In response, the OECD's Chemicals Group developed a set of safety guidelines that would harmonize national requirements and facilitate the growth of a global chemical industry. Beginning in 1985, the OECD Chemicals Testing Programme began issuing standardized protocols for state-of-the-art tests in five areas: physical characterization, ecotoxicity, biopersistence and bioaccumulation, long-term effects on human health, and short-term effects on human health.[59] These became de facto regulatory standards among industrialized countries for commercial chemicals. This framework was intended to sustain and promote the role of chemicals in economic growth, even as it also set in place standards for environmental protection. This is not to dismiss the technical work accomplished by the OECD, which established an infrastructure for chemicals testing, but rather to observe its goal—the continuance of a legacy that privileged and sanctioned chemicals production.[60]

Other kinds of international regulation have become important for chemicals, especially in Europe. In 2006, the European Union introduced a new regulatory regime, the Registration, Evaluation, Authorisation and Restriction of Chemicals (REACH).[61] This law requires registration of all chemicals, except those explicitly exempted, that are manufactured or imported into the European Union.[62] In order to register a chemical, companies must provide data—including that on physicochemical properties, toxicology, and ecotoxicology—to the European Chemicals Agency (ECHA). Many chemical substances had never been subjected to basic toxicology testing, and REACH would require that. Its principle of "no data, no market" departed significantly from the model pursued through TSCA, which obligated the EPA to provide evidence of unreasonable risk in order to require companies to submit data about their chemical products.[63] REACH has been hailed as an example of chemicals regulation that actually implements the precautionary principle—that is, requiring companies to demonstrate the safety of their products before they can be sold.

However, in the past decade, REACH's implementation has disappointed many of those who championed the more rigorous procedures of this law.[64] The degree of oversight depends, in part, on the volume at which the chemical is produced. This means that for commercial chemicals not produced at high tonnage, little data are required. And even for REACH, with its ostensibly precautionary approach for substances that are economically valuable, the process often favors authorization. A study performed by the German Federal Institute for Risk Assessment (BfR) concluded that, for the chemicals used in high quantity (more than one thousand tons per annum), "the examined dossiers were concluded 'non-compliant' in the range of 12 to 61%, depending on the evaluated endpoint."[65] Registration dossiers containing old or merely generic data (especially in regards to the potential uses of the compound) are so uninformative as to be considered vacuous.[66] Under the circumstances, expert committees revert to economic considerations in their scientific deliberations about safety. Even when the relevant data exist, there is a strong "asymmetry of information" available to industry and to regulators. Furthermore, companies have been successful in exploiting various mechanisms of the REACH regulation to obtain authorization for chemicals whose toxicity is undisputed; that are used in large amounts; and that, a priori, should have been banned. The final result is an ostensibly precautionary system that does not diverge so radically from the risk assessment approach of the United States, with its explicit inclusion of economic consequences of regulation and a strong influence of the regulated industries in the whole process.[67]

Environmental nonprofit groups have traditionally leveraged existing laws to strengthen enforcement of national and international regulation, but in the last three decades, other approaches have been developed in light of how much authority chemical industries exercise in the formal regulatory sphere. Corporations and nongovernmental organizations have developed their own regulation in the form of private standards such as the ISO 14000 standards family or the Responsible Care initiative. Correlatively, coregulation, in which public authorities delegate regulatory responsibilities to corporations or to organizations affiliated with trade associations, has become common. Consequently, some activists and NGOs have long chosen to work directly with companies and certain industrial sectors to obtain less polluting production methods and finished products. The phasing out of the use of toxic flame-retardant chemicals in consumer products provides a case in point. The toxic residues of flame retardants are themselves unintended consequences of safety regulations regarding the flammability of textiles.[68] Arlene Blum (founder of Green Science Policy Institute in Berkeley, California) has turned to negotiating directly with large companies on this issue, in addition to working with regulatory agencies at several levels of jurisdiction.[69] In part, the turn to industry is necessitated by the global reach of commercial production and supply chains: China is now the world's largest supplier of chemicals, and Brazil, India, and Russia have large chemical sectors. However, fifty years of patchwork initiatives to transform the chemicals regulation have yet to effectively control hazardous exposures and environmental contamination. One of the main roles of regulatory tools is to create demarcations: between substances, their effects (e.g., hazardous or safe), their legal status (e.g., drug or food), and their use (e.g., at work or in the environment). These distinctions themselves are residual legacies of our use and understanding of the molecular world, foreclosing other ways of envisioning and regulating chemicals.[70]

Waste

On March 11, 2017, on the periphery of Addis Ababa, capital of Ethiopia, a collapsing waste dump killed 113 people and destroyed some thirty houses.[71] This particular waste dump has served the rapidly growing capital's four million inhabitants for over forty years. Most of the victims of the waste pile collapse were locals, workers in an informal economy who, every day, searched for reusable or salable objects among the piles of rubbish. This was not the first disaster of its kind. In July 2000, the landslide of another refuse pile in a slum of 25,000 people on the outskirts of the Philippines' capital, Manila, left an estimated 280 dead and hundreds homeless. Such are the occupational hazards of the "waste pickers" of modern times.[72]

These disastrous moments and the media attention to the funeral processions of their victims highlight, usually just briefly, ignored worlds, the wasteland spaces that are one of the legacies of globalized mass consumption.[73] Sprawling cemeteries of consumer culture's leftovers have risen on every continent save Antarctica (although there is waste there too): the Bordo Poniente dump near Mexico City; the Olusosun landfill in Lagos, Nigeria; the Bantar Gebang in the suburbs of Jakarta, Indonesia; and many others. They are especially visible at the edges of cities in the Global South where they reflect economic growth, galloping urbanization, the incapacity of public service infrastructures, and the inability of local officials to deal with the consequences of rapid, unplanned growth amid expanding inequality. Solutions to accumulating waste can generate their own hazards. Smoke and fine particles from the open-air burning of the garbage mountains in India's capital, New Delhi, contribute significantly to general air pollution and cause many people's deaths.[74]

Northern countries are not immune to these problems.[75] The largest wastelands in the world are located in the United States. Fresh Kills in Staten Island, New York, opened in 1947 and closed in 2001. This site has a surface of 890 hectares (2,200 acres) and retains more than 150 million tons of waste including residues from the Twin Towers that fell on 9/11. The Puente Hills landfill, which began to operate in 1957 and was only closed in 2013, extends over 280 hectares (700 acres) in southeastern Los Angeles, near the city of Whittier, California, reaching 150 meters skyward.[76]

The central place of industrial chemicals in consumer goods, transportation and energy infrastructures, and agriculture determines not only the particular kinds of wastes that litter our planet but also their tempo and scale. From the late nineteenth century onward, industrial innovations with new, often synthetic materials and less costly production processes contributed to the accumulation of more and more leftovers. For waste includes not only garbage (end-use) but also all the by-products of extraction, processing, production, and packaging of goods. Waste is the source of many toxic contaminants in soil and groundwater, such as heavy metals, PCBs, and hydrocarbons. But nontoxic waste is also part of our chemical legacy and contributes its own signature hazards.

Plastics are emblematic of the legacy of large-scale chemical waste and its enduring global impact on the environment. These synthetic compounds are produced in quantities and on time scales that are difficult to fathom: fully half of the 8,300 million metric tons of plastic ever produced was manufactured during the last thirteen years.[77] In turn, plastic is discarded in massive amounts and rates as well, and a significant portion ends up in the seas.[78] Residues are not always small or slow.

For some years now, media reports based on the work of scientists and activists have described what Max Liboiron calls "plastic smog," the Great Pacific Garbage Patch.[79] This immense marine cloud of waste is accumulating in the Pacific Ocean and composed of innumerable bits and pieces originating from plastic bags, bottles, and other floating trash.[80] Several scientific expeditions have studied the phenomenon, and while researchers' attention has focused on the northern Pacific patch of waste, four other equivalent plastic masses have been identified in other waters, and there may be still others not yet identified. These plastic patches form when drifting marine debris accumulates in the middle of a gyre, the clockwise spiral of ocean currents that traps flotsam. The North Pacific subtropical gyre has gathered plastic debris estimated to weigh three million tons and covers an area the size of Texas.

Plastic marine debris can survive several hundred years, disaggregating into microparticles with diameters of fewer than five millimeters that float on the ocean surface and in the water column down to thirty meters. Animals ingesting these microparticles cannot digest them and often have difficulties eliminating them. Plastic obstructions can cause death. But even when not so immediately hazardous, these particles accumulate in the bodies of fish, tortoises, birds, and other marine animals. Most plastic polymers are sponges for hydrophobic toxic chemicals, including persistent organic pollutants such as DDT or PCBs. Thus ingested plastic releases a toxic load of contaminants in marine life, which in turn can concentrate up the food chain. In sum, plastic waste has major consequences for the environment and, through seafood fished from the ocean, for human health.

This situation is partly a function of the global scale and the voracity of consumer capitalism. But it is also a legacy of past decisions and institutional patterns of managing waste. Today, different modes of waste management tend to correspond to three distinct logics: a logic of containment, exemplified by "controlled" storage of waste in landfills or lagoons; a logic of dispersal as when waste is dumped into rivers, lakes, and seas or when debris is reincorporated into the earth as "fill" for landscaping or infrastructure development projects; and a logic of treatment or processing, illustrated by a range of technologies involving incineration, recycling, or reuse. Another feature of waste management is its tendency to reinforce social, racial, and environmental inequalities.[81] Illegal waste dumping in minority communities and the unequal toxic burden accompanying the practice were driving motivations for environmental justice movements in the United States nearly fifty years ago and remain some of the vehicles for its internationalization today.[82]

Over the years, different combinations of waste management solutions based on these different logics have been promoted and implemented by government regulators, citizens' groups, social movements, and industry representatives. So far, however, none have halted or even appreciably slowed the mounting volume of waste, which all too often simply overpowers the technical capacities of states and industries to adequately manage this legacy of residues.

The invention of the modern concept of waste coincided with a transition from domestic garbage management based on do-it-yourself disposal and reuse toward the emergence of municipal and national policies of waste management.[83] As Sabine Barles shows in her analysis of Parisian waste management, this transformation followed in step with industrialization and corresponded with a transition in how raw materials circulated between cities and rural areas, as well as within cities. As in other industrializing cities, the Parisian garbage provided raw materials for other production sectors: rags for paper production, bones for combs and buttons, and excrement for compost manufacturing and biological fertilizers used in agriculture.[84] To be clear, industrializing cities of the mid-nineteenth century had waste problems. Municipal public health offices came into existence in part to deal with the diseases resulting from inadequate disposal of human and animal waste. However, the rise of the chemical industry meant that repurposed castoffs were gradually replaced by inorganic mineral resources (phosphate, potassium hydroxide) or by synthetic materials created by chemists, with cellulose replacing rags or resins substituting for bones.

From the late nineteenth century onward, this new chemical economy meant that postconsumer waste became a different kind of problem for cities to manage. Since garbage was no longer useful, it had to be somehow eliminated. Incineration was considered to be the most hygienic solution for waste treatment, but burning waste in large quantities was expensive and proved to be insufficient to cope with growing amounts of trash. Waste began to be placed in dumps, initially as a holding bin for later incineration but increasingly as a form of disposal, buried or piled. Throughout the twentieth century, trash accumulated at a growing number of dump sites outside of cities. In the United Kingdom, "controlled" waste storage in "sanitary landfills" was promoted beginning in the early twentieth century. However, throughout the century, informal, unregulated dump sites outnumbered the legal, regulated ones. Seas, rivers, and lakes have long been the sink of choice for hazardous wastes.[85] Discoveries of toxic industrial waste in Love Canal, New York, and from the incomplete cleanup of the chemical accident in Seveso, Italy, made the news, whereas the ecological consequence and health burdens of most waste were unnoticed.[86]

The blossoming environmental awareness of the 1960s played a decisive role in changing policies and practices of waste management and contributed to a new discourse and understanding of waste as a public problem for which the waste management industry positioned itself as an eager problem solver. In 1965, the Johnson Administration issued the Solid Waste Disposal Act.[87] Justifying this bill were alarming pictures of the waste situation: aging and saturated incinerators, open-air used car dumps in the suburbs, and mountains of waste rock around mining sites. In 1975, two-thirds of household waste ended up in the environment, particularly in rivers and lakes or in illegal dumps.[88] In his study of garbage in the United States, Martin Melosi asserts that the U.S. Coast Guard supervised 120 coastal sites as legal waste dumping areas.[89] In Tokyo and New York, offshore waste burial was incorporated into land reclamation projects, in which "land" was reclaimed from the sea for urban expansion.

Since the 1970s, a key challenge of waste management has been to bring the long-term environmental costs of used products into the economic cycle, such as through the "polluter pays" principle. The policy of extended producer responsibility (EPR) promoted by OECD obliges manufacturers to reprocess the waste their products ultimately generate.[90] In Europe in recent decades, more than twenty manufacturing sectors have been established to manage their waste on the basis of the EPR, among them electric and electronic waste management. Implementing these new waste policies means dealing with the multinational companies that have taken over waste management such as American Waste Management Incorporated, a world leader in waste management, or the French *Véolia Environnement* and *Suez-Sita*.[91]

For some years now, waste has become relevant to the wider discussion of a green economy and energy transition. Growing environmental concerns related to energy transition and a green economy coupled with related concerns about securing access to strategic raw materials have gradually transformed the status of waste. Today, waste is more and more frequently promoted for its economic value as "green gold" or, in some cases, an "urban mine"— that is, a potential supply of much-needed resources, newly accessible through advanced techniques of recycling and valuation. Even so, the legacy of waste is not easy to redeem. Problems related to the sheer amount of waste and its potential toxicity persist, unresolved. Waste management scandals appear regularly and highlight the weaknesses of regulation, occasionally revealing corruption and the diversion of waste into organized crime in the United States and Italy or in the global trade of toxics.[92] More than that, hazardous waste has grown to such an extent that it worries the public authorities: the Council of the European Union, for example, has selected it as one of the European Union's top ten priorities in the fight against organized crime for the next five years.[93]

Waste remains a burdensome legacy to be managed both materially and politically. Seen from a *longue durée* perspective, "waste management" has helped to legitimize the continued accumulation of waste, buried in closed dumps and terraformed landscapes or shipped from wealthy northern nations to poorer regions in the South. In this sense, policies and practices meant to control waste through rational management are also implicated in the production of its continued invisibility. The Fresh Kills landfill is on its way to becoming Staten Island's largest city park, targeted for completion by 2037.[94] This park will cover waste with greenery and water features, thus symbolizing the rehabilitation of the world's largest waste heap. But the residues within the park will not have disappeared. They will remain—sequestered and monitored, but not gone.

Conclusion

This chapter has explored many legacies of chemical residues. They include communities making sense of their past and their future, regulatory regimes that are multilayered and recalcitrant to change, and the masses of consumer and industrial waste that litter the globe, as well as the actual dispersed environmental contaminants that have put toxic traces of human activity even in the wildest places on earth. Chemical residues are often unseen, which makes it easy to forget the past. Paying attention to the irreversibility of residues, the path-dependence of contamination and regulation, and the durability of industrial leftovers serves as a sobering antiamnesiac.

By focusing on legacy as a key theme in understanding chemical environments, we have sought to illuminate the material, social, and administrative connections between the past, the present, and the future. There are numerous ways in which states, communities, and individuals respond to residues—to contain, manage, or even disguise their presence. And some residues are less troublesome than others, to be sure. But the chemical rearrangement of the world, which has accelerated in the twentieth century, is not reversible, and even those residues that are ostensibly nontoxic, such as many plastics, can threaten ecosystems through their physical mass and interactions with other pollutants.

To the degree that we can also think about environmental regulations as residual and accretive, their durability can create problems as well. Decision-making procedures (such as toxicological testing or risk assessment) may be impervious to new knowledge, and the implementation of regulations may be out of sync with the preexistence or persistence of pollution. Most chemical regulations are prospective and do not address environmental contaminants that have already accumulated or those that will accumulate after a product is disposed of. The growing emphasis on "extended producer responsibility," in which manufacturers reprocess waste from their products, has come from the OECD, which lacks the power to enforce its guidance. It remains to be seen whether industry self-regulation will be effective in better managing the problems of toxicity associated with consumer waste.

And of course, consumer waste is only one source of contaminating residues. Industrial waste has been accumulating in the environment for centuries, though the volume has increased hugely through the mass production of chemicals since the nineteenth century. It is often governments, rather than companies, who are on the hook to clean up waste sites after the fact. Such environmental remediation is driven by a different set of laws, techniques, and obligations than prospective chemicals regulation. In the United States, the much-publicized Love Canal tragedy, in which buried hazardous waste leached into a neighborhood built on top of the site, prompted Congress to pass the Comprehensive Environmental Response, Compensation, and Liability Act of 1980 (CERCLA). This law charged EPA with administering a cleanup program of highly contaminated hazardous waste in what were termed "Superfund" sites. However, the bill's original strict liability scheme (introduced by Congressman Al Gore) was weakened over the course of negotiations with the Senate, meaning that just determining culpability has become a costly legal exercise.[95] Moreover, when residues transition from being worthless externalities to waste that must be remediated, they become exceedingly expensive. In this sense, residues have a bizarre economic value, directly proportional to their perceived danger.

Needless to say, the costs of environmental contamination are not only monetary. When it comes to disease burden, the legacies of residues are unevenly distributed in ways that overlap profoundly with social inequities, especially those of race and class. Attention to residues underlines the urgency of work to document and understand environmental injustice while pointing to the importance of paying attention to every step of the lifecycle of industrial products, from hazardous waste generated by manufacturing plants to the garbage mounds of Ethiopia, the Philippines, Mexico, Nigeria, and countless other places. Health consequences of exposure to toxic residues, especially those at low doses, can take decades to manifest. And when the impact is an increase in the incidence of common chronic health problems such as cancer and heart disease, the role of chemical exposures is hard to discern, and even harder to document sufficiently for the purposes of legal redress.[96] The legacies of many residues are both long-lived and imperceptible.

The durability of residues intersects with other temporal frames, such as the increased pace of "natural" disasters in an era of anthropogenic climate change. Already in the 1970s, geographers Ian Burton, Robert Kates, and Gilbert White pointed to the role of "second nature," the technological systems at the boundary of water and land, or urban and undeveloped space in enabling modern disasters to occur.[97] In turn, chemical contamination poses one of the greatest long-term dangers of both industrial accidents and weather events such as Hurricane Katrina.[98] Like residues, disasters do not respect the difference between nature and technology.[99] Drawing on Rob Nixon's idea of "slow violence," Scott Knowles has gone so far as to speak of the unremitting pace of "slow disaster": a concept that resonates with our attention to the chemical legacy of residues.

Changing course with respect to the current production, consumption, and disposal of chemicals is incredibly difficult, economically, politically, socially, and materially. In the end, we are not trying to suggest that these legacies of contamination are fated. Legacy is not destiny, but we cannot expect markets alone to reorient chemicals production toward less environmental harm. Moreover, the challenges of residues are not only about their accumulation over time but about their spreading over space and cycling through ecosystems. Charting the sometimes-haphazard geographical reach of contamination is our next step.

3

Accretion

Before the mid-twentieth century, industrial use of radioactive materials was tied to particular places and types of employment. As a result, the scope of damage it caused to environments and human health was relatively contained. Uranium mining began in Europe in the late eighteenth century. Glassmakers and ceramists used the yellow ore as a coloring agent, its primary industrial use until 1898, when French scientist Marie Curie famously discovered radium's special property of radioactivity and opened a new era of applications for naturally occurring radioelements. But even as industrial and medical uses for processed uranium expanded, production of such radioelements remained place-based well into the twentieth century, and health risks from radiation exposures were mainly occupational. Uranium miners, women radium dial painters, and scientists were unwittingly exposed to the now well-known biological risks.[1] But by today's standards, radioactive residues did not travel far beyond these occupational groups, mines, or factories; most naturally occurring radiation remained encased in unmined deposits belowground and out of harm's way.

Since 1945, triggered by the events of World War II, artificial radioactive elements (radioisotopes) have permeated the planet. Industrial, military, and medical uses have multiplied. Processed forms of plutonium, radioactive cesium and strontium, tritium (radioactive hydrogen), and carbon-14 have proliferated. The 528 aboveground atomic weapons tests that took place from 1945 to 1964 (when they were banned) disseminated the products of nuclear fission around the globe, some with half-lives approaching eighty million years (plutonium-244). Geochemists have registered the unique atomic signatures of these radioelements in rocks, soils, lake sediments, clouds, and plants; medical scientists have identified them in human bones and babies' teeth. Their presence offers irrefutable stratigraphic evidence—a "golden spike"—for the Anthropocene: a human-scale production that took decades but whose radioactive by-products will endure for epochs, in terms of earth's geological history.[2]

Today's global budget of radioisotope pollution is an irreversible legacy of twentieth-century science and technology. We use the example to open this chapter, on accretion, as it also illustrates how the concept of residue functions as *a dual process of accumulation and diffusion*, gathering over time and building up across space. If legacy signals a tendency toward persistence, accretion asks us also to think about spatial mobility and the rhythms of change. The distinction is an important one. To understand why measurable levels of artificial radioisotopes have come to blanket the Earth, one needs to pay close attention to the intricate histories of nuclear weapons, nuclear energy, and nuclear disaster—histories that designate particular times (1945), places (Bikini Atoll), organizations (the International Atomic Energy Agency), and events (mass protests) as critical to charting radiations' spread. Radiation also changes the places it inhabits. The new sciences that grew up in the atomic bomb's wake, such as radiation biology and radioecology, have documented the erratic behavior of fugitive radioelements, observable through earth geochemistry, ecosystems ecology, and human health. When radioactive isotopes decay, they often transmute into other chemical elements, but in addition, a radiomaterial may take different forms as it is unearthed, used, and dispensed with. Uranium ore is drawn from mines, pressed into yellowcake, and enriched into uranium-235; the rest is discarded as tailings. Loaded into reactors, this fissile uranium-235 releases the energy that buzzes along electrical wires while also generating radioactive waste. But when uranium is cast off as mine waste or released accidentally, it becomes dust picked up by the wind and settles in the lungs of downwinders, sometimes inducing cancer.[3] Following this radioactive elements' twisting trail, we find that accretion is not a random process. Its paths are circuitous but also structured and decidedly nonuniform.[4] When residues travel, they spread out—unevenly—to delineate patterns of social power and disadvantage.

Accretion refers to the gradual accumulation of a substance, often growing through processes of surface deposition. We use it here to emphasize not only how residues move in space, their geographical and topological reach, but also how small-scale and imperceptible those movements may be at any point in time. Residues disperse, but they also stack up, sometimes unnoticed until they become a problem. Three cases help us map the dynamic twists and turns of residue accretion. The first case excavates a small urban lot to uncover the regulatory processes leading to the discovery and remediation of an underground gas leak, the second tracks asbestos fibers circulating through the building and construction trades as they become woven invisibly into urban environments, and the third follows rare earth elements as global markets route them and a cocktail of accompanying toxics along international commodity chains. In its own way, each story illuminates mundane and often taken-for-granted processes that incrementally but continuously reorganize and redistribute the chemical leftovers of modern (and not-so-modern) life across the planet. They also speak to the social, biological, and ecological consequences of the circulatory systems that set residues in motion.

Of the five residue properties explored in this book, slipperiness gets special attention here because it most clearly evokes the dynamics we intend to measure with the term *accretion*. The substances we follow are escape artists. They move around a lot, mostly undetected, often out of sight. One hides underground. Another hitchhikes. And the third performs tricks of transfiguration. Their routes are determined in part by their own physical properties, whether organic or inorganic, liquid or solid, volatile or inert. But even more interesting for us is how the patterns of accretion are determined through interactions with nature. In the stories that follow, weather, landscape, soil structure, and hydrology matter a great deal in understanding the social and ecological forces that keep residues on the move. In the end, however, economic interests work hard to screen these residues off as much as politically possible, keeping them out of sight, sometimes on account of the slow pace of accretion itself. In other instances, and somewhat ironically, traditional safety standards are responsible for continued use and wrongful deposition of harmful substances. As matter out of place, and often on the move, residues erode our regulatory systems in unforeseeable ways. Our focus on accretion is meant to coax residues out of hiding and into the light of day.

Gasoline

The Perry and Marty Granoff Center for the Creative Arts (GCCA) cuts a distinctively modernist profile on the campus of Brown University in Providence, Rhode Island. Completed in 2009, the four-story building boasts a sharply angular design featuring a glass, metal, and concrete exterior. The interior features high ceilinged galleries as well as studio and classroom spaces and offices for Arts faculty. Outside, a wide concrete patio spills onto a small but tastefully landscaped green, an inviting spot for students and other passersby to sit and relax during warmer months (see figure 3.1, left photo).

The site's land use history is considerably less distinctive (figure 3.1, right photo). For the first half of the twentieth century, two houses occupied the half-acre site. In the early 1950s, the lot was converted from residential to commercial space and became home to Colbea's East Side Service Station, a locally owned gas station and auto service garage that served the campus neighborhood's growing population of midcentury motorists. Under different names and leasing arrangements, the site continued its life as a gas and auto service station for the next five decades until it was demolished in 2006 to make way for the GCCA.[5] This latest land use transition was not without complications, and it is the complications that provide fresh insight into some of the ways

FIG. 3.1. GCCA Site Photos. (Left) foreground: Perry and Marty Granoff Center for the Creative Arts, 2017. Photo credit: Scott Frickel. (Right) Colbea Service Station, circa 2004. Photo credit: GZA Environmental, Inc. 2006. *Site Investigation Report: East Side Service Center*. Providence, Rhode Island.

that residues are slippery, regularly escaping engineered systems to accumulate in out-of-the-way places.

Gas station engineering requires a subterranean infrastructure assembled from large underground storage tanks (UST) and from lines connecting the tanks to one another and to a series of above-ground pumps and meters. On the GCCA site, ten USTs were in commercial use at one time or another. Early on, these included three 6,000-gallon gasoline tanks, a 550-gallon waste oil tank, and a 550-gallon fuel oil tank. In the late 1980s, some of these USTs were removed to make room for three much larger (12,000-gallon) gasoline tanks and a new 1,000-gallon waste oil tank.[6] Public knowledge about this privatized underground infrastructure remained buried as well until the fall of 2006, when occupants of an adjacent campus building reported strong chemical odors inside the building's ground floor. Investigation of the building's subbasement drainage system along with water testing from newly drilled monitoring wells revealed the presence of a layer of gasoline and constituent chemicals floating on the groundwater table.[7] It was only after this discovery that the gas station's manager reported an accidental release of gasoline believed to have originated from an earlier break in one of the buried lines connecting two USTs. Reports suggest that the line break had flushed an unknown quantity of gasoline into the subsurface soil, glacial till, and fractured bedrock; the resulting gas plume had migrated through the aquifer, moving down-gradient across the gas station lot and underneath the building next door. It was also moving beneath the street and would eventually make its way onto campus lots on the other side.

To investigate the gas leak, technicians monitored contaminant concentrations in groundwater and in the office building's drainage and air ventilation systems. They also assessed subsurface hydrogeological conditions; evaluated the size, composition, and flow rate of the contaminant plume; and implemented a complex remediation plan. This plan included, but was not limited to, excavating and removing 1,600 cubic yards of gas-impacted soil, injecting 1,000 pounds of a commercial chemical oxidant into shallow soils in an attempt to neutralize remaining volatile organic compounds, and installing an activated carbon filtration system to treat groundwater entering the new GCCA's drainage system.[8] Beyond the gas leak, the regulatory process also included standard EPA-mandated Phase I and Phase II site assessments as well as further characterization of nongas soil contaminants (arsenic, lead, and mercury) and polycyclic aromatic hydrocarbons that had been identified from chemical analysis of Phase II soil samples.[9] Environmental consultants leading the site assessment then developed a second remediation plan to deal with the high levels of arsenic and other contaminants not related to the gas leak. This second plan combined soil encapsulation methods with restricted site-use protocols, the latter to help ensure that the now-capped soils are not disturbed by future uses, including new construction projects, that might reexpose contaminated soils to the ambient environment and again threaten public health.

The gas leak was a major wrinkle in an otherwise fairly standard environmental assessment. Begun in 2005, regulatory procedures conducted on and around the GCCA site continued for another four years. In full, the environmental site assessment involved considerable sustained effort conducted by dozens of people and organizations. Their work is preserved in a documentary tangle of plans, reports, addenda, decision letters, photographs, maps, and test results. Through all this, a diverse group of environmental specialists, workers, and campus and regulatory agency officials produced, reviewed, and certified a cascade of locally contingent knowledge focused on a single, mundane kind of place: an old gas station. In turn, this knowledge opens a much larger window on a more universal set of practices and processes through which chemical residues thicken up and thin out.

Residues thicken up when they accumulate in particular places, as when a burst line sloshes hundreds of gallons of liquid fuel beneath the pavement of a neighborhood gas station. Residues thin out when they spread to other places: in this case, by migrating downhill through an aquifer and into a nearby office building. But there are many ways that accretion works with residues—thick and thin, fast and slow, regular and sporadic, intentional and accidental, acknowledged or not.

In many ways, the story of the GCCA site is utterly ordinary. Soil and groundwater contamination are logical—if not inevitable—consequences for any business, small or large, whose practice involves storing large volumes of highly toxic fluids underground for decades at a time. U.S. federal law recognizes these risks, and regulators have developed detailed inspection, reporting, and response protocols for dealing with USTs, protocols that were yoked into action at the GCCA site.[10] Still, our study of residue accretion on a single urban parcel, small though it is, raises larger questions about the capacity of regulatory agencies and other actors to identify hazards, assess risks, and mitigate the environmental and health consequences of chemical contamination.[11] How and under what conditions does regulatory science operate to forestall the accretion of chemical residues in particular places?

Posing the question in this way points to elements of the case that are perhaps not so ordinary after all. Notably, its location on a densely populated urban university campus meant that ignoring the problem was not a practical option for university officials, especially since university employees and students occupying an adjacent building initiated the discovery. Overnight, the site became a potentially significant public relations problem that threatened progress on a hallmark new building; the environmental conditions on the site demanded swift attention. In addition, the university's status as a private, not-for-profit educational institution, with a significant endowment, meant that the financial resources to undertake the costly remediation were within reach. It matters, too, that Brown University enjoys a good reputation among state regulatory officials as an environmentally enlightened property owner, albeit one that is also eager to avoid potential fines and lawsuits related to noncompliance with environmental laws. University administrators and legal counsel take seriously their statutory obligations to investigate potential risks on its many properties, both on and off campus, even when it means going beyond the strict letter of legal compliance. Brown is one of the city's larger property owners, so the university's commitment to environmental best practices has broader, substantive consequences for local regulatory compliance. The university's sizeable endowment makes it unusually sensitive to liability concerns, after all. For these reasons, the GCCA site is almost certainly atypical, but in ways that help, not hinder, thinking through chemical environments: compared to the two other examples that we use in this chapter—asbestos and rare earth minerals—this one offers a "best case" for exploring the limits of what (U.S.) regulatory practices can (and should) actually accomplish when the problem of accumulated chemical residue is taken seriously and pursued responsibly as a matter of sound fiscal investment.

When conducted conscientiously, site assessments generate reams of new data describing the sites' land use history, ecological and hydrological dynamics, contaminant levels, possible exposure routes, and potential risks. This information is compiled in reports and used by regulators to decide whether and how to deal with chemical residues. Yet while government offices continue to provide oversight, state officials today participate less and less centrally in the work of regulation. Instead, much of the knowledge about contamination is produced, organized, and circulated in private or quasi-private domains by property owners, real estate developers, construction companies, law firms, environmental consulting firms, engineering companies, and analytical laboratories. How this industry coordinates the efforts to meet regulatory requirements for individual parcels of land is not well understood because there has been very little research focused on how knowledge developed from site investigations is actually produced and where it circulates. So while the nature of this knowledge makes it consequential, it is less clear if the knowledge is durable, much less accessible.

Analysis of documents related to the GCCA site also suggests that "successful" regulatory site assessment and remediation does not mean that chemical residues are no longer present. Few remediation strategies actually involve removing all contaminated material and replacing the soil with clean fill. Far more often, remediation is in situ. In recent years, "monitored natural attenuation" has emerged as an increasingly widespread remediation strategy.[12] By letting nature take its course through phyto- or biodegradation processes, natural attenuation can take time—years or even decades in some cases. But as long as engineered controls keep contaminants contained on-site and away from people, doing nothing is an attractive option—and cost-effective. At the GCCA site, notwithstanding the 1,600 cubic yards of gas-impacted soil that was dug up and carted off, remnants of the gas plume still lurk below the surface even as a new system of retaining walls, trenches, ventilation systems, geotextile capping material, and landscaping prevent it from moving too far, too fast, or too close to people. In combination, these remedial controls seem to work pretty well, except when they don't.[13]

A related point is that in the United States, most postmarket regulation of chemical residues is site-based and follows an economic logic organized by the principle of private property. Under provisions of federal law, contaminated land is the responsibility of the property owner, even if the contamination originates somewhere else. Owners of urban parcels are required by law to report accidents that result in hazardous releases.[14] Otherwise, only when the property enters a real estate market are site assessments conducted, generally as a condition of sale. This means that, in general, site-based knowledge production follows the rhythms of real estate markets—more frequent when markets are hot, less frequent when they are cool.[15] One problem with this way of organizing regulation is that contaminants often move around, transgressing property lines. It is easy to forget that urban systems are also ecosystems and nature seldom abides by the conventions of property law. Wind, rain, gravity, erosion, and even the structure and composition of subsurface soils influence how contaminants are mobilized and where they go. Property lines and the associated legal responsibilities are of no concern to a chemical plume that is moving downhill.

The mismatch between the property logic of regulation and the meandering illogic of residues has a scale dimension as well. Whereas sites are regulated and remediated one at a time, on a case-by-case basis, the contaminants they are targeting often move through entire neighborhoods or even regions, if in nonuniform ways. Focusing on the accretion of residues illustrates the basic conundrum that law is a rather poor tool for dealing with environmental problems. The dilemma is especially evident in the United States, with its valorization of private property (above all the single-family home), but even a sturdier environmental policy attuned to public goods may not be able to reckon with the wiliness of residues on the loose.

There are an estimated 145,000 to 150,000 privately operated fueling stations pumping gas in the United States today. If this number seems large, consider that it represents a steep decline from 1994, when the number of retail fueling stations hovered around 203,000.[16] In Providence, where the drop has been much steeper, today just 28 gas stations service area residents. Yet analysis of past issues of the *Providence City Directory* identifies 526 gas and oil service stations that operated in the city between 1936 and 1990 (see figure 3.2).[17] The city's gas station population actually peaked in the 1950s, well before government agencies began maintaining underground storage tank databases.

In this context, it is worth asking whether the gas station leak at the GCCA site is distinctive, not because of the subterranean contamination that created all sorts of headaches for University

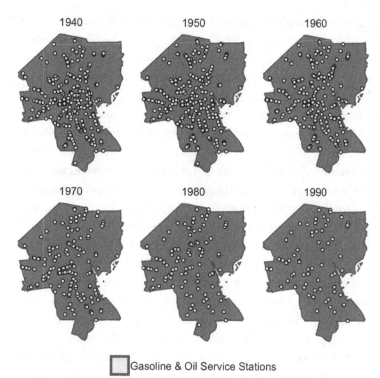

1940   1950   1960

1970   1980   1990

☐ Gasoline & Oil Service Stations

FIG. 3.2. Gasoline and Oil Service Stations, Providence, Rhode Island, 1940–1990.
Source: *Providence City Directory.* Map credit: Thomas Marlow.

administration and those in charge of redeveloping the property, but in the sustained and apparently comprehensive nature of the regulatory response. What then of the roughly eighty thousand sites across the United States once occupied by gas stations?[18] What, if anything, is known about those former gas station lots? And of course, it is not just gas stations that should concern us. Other types of sites store, sell, and likely leak large quantities of hazardous materials into the earth—dry cleaners, auto repair shops, paint stores, and pest control businesses among them. And then there are industrial manufacturing facilities.

In heavily industrialized urban areas of the United States, land use change plays an important role in spreading, accumulating, and hiding industrial hazards.[19] A key process is industrial churning, or the regular movement of industrial activities as factories open, operate, close, and move away. Industrial churning is a basic feature of urbanization and is fundamental to the economic health of cities large and small. Most importantly, industrial churning never stops. Instead, industrial churning operates continuously with the result of spreading chemical residues throughout urban areas—from site to site, year after year—altering the chemical composition of urban soils, water, and air.

This type of residue accretion is mostly invisible. Endemic resource shortages in federal, state, and municipal regulatory agencies virtually ensure that only the sites that are the largest, visibly dangerous, or the most economically viable for redevelopment will attract regulatory attention. The vast majority of other sites are effectively ignored: the smaller businesses, the shorter-lived operations, the plants that do not attract negative attention from community members. These types of places, regularly bypassed by regulatory oversight, account for about 90 percent of historically existing industrial parcels.[20]

In urbanized Rhode Island, the macroview of residue accumulation created by industrial land use change is sobering. Since 1950, the small state's footprint of industrialized land has grown continuously to include nearly seven thousand unique parcels, almost all of them (88 percent) clustered in metropolitan Providence.[21] Figure 3.3 registers the result of industrial churning visually. Moreover, this accumulation has occurred unrelentingly, even throughout a two-decade period marked by intense deindustrialization (roughly 1980–2000), as measured by employment in manufacturing.[22] By contrast, the EPA's Toxic Release Inventory database for Rhode Island includes just 404 sites that have reported releases since the database was established in 1987. Here is the key point we want to emphasize: the same regulatory system that provided extensive oversight on the GCCA site on Brown University's campus has conducted no regulatory oversight on 94 percent of current and former manufacturing facilities that have operated in Rhode Island since the 1950s.

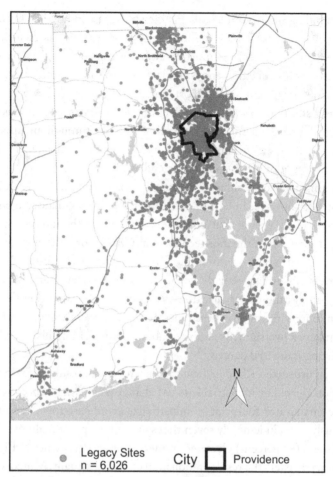

FIG. 3.3. Legacy manufacturing sites in Rhode Island, cumulative
1953–2016.
Data source: *Rhode Island Manufacturing Directory*, various dates. Map
credit: Thomas Marlow.

As long as industrial activities churn across cities, residues like
gasoline—or asbestos, which we encounter next—will accumulate
even as they become lost in the shuffle of urban redevelopment. And
as long as regulatory agencies do exactly what they are required to do
by law and nothing more, those lost sites will likely remain lost, and
the accretion of chemical residues will continue, more or less unabated.

Asbestos

We learned in the previous chapter that "asbestos built" Ambler, but as a residue, its reach into built environments extends far beyond the little Pennsylvania town. Like radiation, asbestos is emblematic of the manufactured and largely invisible risks of the twentieth- and twenty-first-century technologies that sociologist Kai Erikson once dubbed "a new species of trouble."[23] But whereas radiation trains our focus on the patterns of accretion shaped by the geopolitics of energy and war, the spread of asbestos has a more domestic pattern, emanating from our homes, schools, and offices and involving the building and construction industries that design and create them. As a central ingredient of modern construction technologies, asbestos is literally molded into the brick-and-mortar edifice of modern daily life. But encapsulation and "controlled use" practices do not always protect against asbestos exposure. Like the gasoline escaped from underground storage tanks to seep into soil, asbestos is also subject to natural forces, swept here and there by wind and rain. But unlike our gasoline example, asbestos also follows specifically social paths, leaving industrial spaces as dust on the clothes of asbestos factory workers making their way home at night.

Chemists and engineers have long considered asbestos a "miracle" product owing to its strength, flexibility, and durability, as well as for its insulating and heat- and flame-resistant properties—all of which have led to the development of a wide range of applications, from firefighting suits to asbestos-cement, including spray insulation for ships and buildings. Asbestos is a fibrous silicate mineral. Crushed into a powder-like substance, its particles can reduce to microscopic size. Each gram of asbestos may contain several thousands of individual fibers, and these can travel invisibly as dust, wherever and as far as the wind, rain, and people can carry it. Under a microscope, asbestos fibers look like ragged needles, and it is this shape, combined with their microscopic size, that makes the miracle mineral especially dangerous. Inhaled or swallowed, the fibers are easily lodged in the soft tissues of throats, lungs, and pleura, where their deadly work occurs imperceptibly over decades and often without advanced warning.[24]

The damage asbestos does to bodies is well documented. In the early twentieth century, biomedical scientists showed that asbestos can cause asbestosis, a disease similar to silicosis. In the 1950s, scientists learned that asbestos exposure is associated with lung cancer, and in the 1960s, they discovered mesothelioma, the signature cancer of asbestos—that is, a cancer whose only known cause is asbestos exposure. But what makes asbestos especially insidious is that exposure to even very small quantities can produce significant human health effects. Although the higher the exposure, the greater the risk, a single fiber sunk into a vulnerable cluster of cells can trigger a disease process that may end—years or decades later—in fatality.[25] The number of deaths from asbestos-related diseases is staggering, as are the estimated costs of compensation to disease victims, although perhaps not so surprising given production trends.[26]

Worldwide, more than 200 million metric tons of asbestos have been produced since 1900, most of it mined from just a few countries.[27] For perspective, France's share of this total translates to about 80 kilograms (176 pounds) per person. Countries such as China and Russia have been consuming more than 500,000 metric tons of asbestos per year since the early 2000s, trends that show signs of slowing only from the mid-2010s. Most strikingly, more than 80 percent of total asbestos came into use *after 1960*—that is, after scientific confirmation of asbestos's carcinogenicity and confirmation of its serious risks to public health.[28] Because asbestos is virtually indestructible—another of its many "miracle-making" properties—the continued mining and use of asbestos constitutes a major public health issue, albeit one that is far less visible in public than, say, fatalities from smoking or automobile accidents.

Unlike the gas station example above—which we presented as a "best case" illustration of site investigation and remediation—in many ways, asbestos illustrates the opposite. At least as judged by the level of public contestation and public health consequences, decision-makers have not taken the clear and widespread dangers of asbestos residues sufficiently into account.[29] Regulation has been slow to develop, and policies vary considerably from country to country. Today asbestos use is banned in fifty countries, including the twenty-seven states of the European Union, but it is permitted in India, China, Russia, Brazil, and Indonesia, to give just a few examples of the main asbestos-using countries. Even in the United States, asbestos is still authorized, though its uses are regulated. In 1989, the EPA issued an Asbestos Ban and Phase-Out Rule, but the rule was successfully challenged in court by industry.[30] Even in the countries with the most protective regulations, laws governing asbestos differ significantly depending on whether they cover environmental or occupational exposures. In many countries, paradoxically, workers are exposed to much higher levels than the general population but are entitled to less compensation than victims of environmental exposure.[31]

Shipyards, insulation factories, and asbestos processing plants attracted the most concern from regulators early on because workers in those industries were routinely exposed to asbestos in very high doses. More recently, in countries that already banned or restricted asbestos use, regulatory agencies and public health experts have expanded their purview to the numerous industries in which work-related exposure to asbestos may be indirect or more occasional. Nowadays, construction workers are the largest group of occupationally asbestos-exposed workers, and they constitute a significant proportion of people suffering from asbestos-related mesothelioma, bronchopulmonary cancer, and other diseases.[32] It's not difficult to understand why. At construction sites, plumbers, roofers, electricians, and others are exposed daily to materials containing asbestos, from cement reinforced with asbestos fibers to different materials used to protect, decorate, or reinforce floors, walls, and ceilings. They may be exposed to high doses of asbestos when dismantling sinks, pulling electrical wires through walls, or changing roof tiles.

Yet most exposure risks remain invisible. Knowledge about which products contain asbestos is complicated to maintain; there are no inventories of such products or registries of exposure risks, and most workers are never informed about this aspect of the material conditions of their labor. The same lack of knowledge that renders asbestos exposures hard to avoid means that most exposures go unobserved, unrecorded, or at least underappreciated.[33] Protecting workers from asbestos is similarly complicated and uncertain. Since the potential presence of asbestos at a given site is impossible to evaluate, a truly protective strategy would require all workers to always use extremely restrictive (and expensive) equipment—full-body suits, head coverings, masks, breathing apparatuses—which is expensive, uncomfortable, and impractical. So, typically, very little is done to protect workers on the job. The nonmanagement of this situation adds to the limited visibility of these exposures and to the widespread but very mistaken impression in many countries that asbestos exposures are declining or gone, a part of history: a legacy built on a lie. The contrasts between the regulatory response to a gasoline leak in our first example and the nonresponse or nonmanagement of asbestos exposure on many construction sites are striking too. The differences highlight the strong discrepancies—and artifice—between environmental and occupational regulations.

One explanation for the lack of regulation has been the persistence of scientific controversies promulgated by the asbestos industry concerning asbestos exposure and health risks.[34] As in the case of tobacco or leaded gasoline, justification for continued manufacture and use of an extremely dangerous product involves a set of very specific strategies.[35] In this case, the main strategies have involved differentiating asbestos into more and less dangerous types. For example, industry has mobilized studies to advance the view that chrysotile, the most commonly used type of asbestos, is less dangerous than other types.[36] (This view is not backed by a strong scientific consensus; the World Health Organization's International Agency for Research on Cancer classifies all forms of asbestos as carcinogenic.) Industry and allied scientists also assert that exposures to chrysotile are safe below a certain threshold. Trade and legal representatives of the industry have used the two strategies together to uphold the view that workers can safely engage in the "controlled use" of asbestos.[37] More recently, a controversy initiated within the auto repair industry, whose workers are exposed to asbestos contained in vehicle brake pads, has centered on a related strategy of differentiation, promoting the claim that short asbestos fibers are less dangerous than longer fibers, the only ones older microscopes were able to detect.[38]

Strategies for continuing use of asbestos have succeeded in relativizing the hazards of the main occupational carcinogen, yet it's one that national governments and various regulatory agencies are hesitant to eliminate. The rise of risk assessment and risk management as mechanisms for environmental administration has also been instrumental in legitimizing the continued legal use of asbestos by giving economic considerations an explicit role in regulatory deliberation.[39] In turn, decisions authorizing continued asbestos production and use have played an important role in the dynamics of its accretion.

One of the main regulatory tools that has helped asbestos manufacturers promote controlled use of asbestos is "limit values." The use of limit values to protect workers presumes that a physiologically tolerable level of a substance exists, below which either it is safe or the risk is so negligible as to be considered socially acceptable, usually for economic reasons. The "limit value" or "threshold limit value" represents that specified dose for the particular substance and a specified mode of exposure (e.g., ingestion or inhalation).[40] Often the population exposed is also specified. In a workplace, limit values function as safety thresholds or pragmatic exposure levels not to be exceeded. They have been used for decades to help companies and administrations manage potentially hazardous situations. Since the 1970s, regulators have regularly updated limit values for asbestos; these have made it politically possible to manage both occupational exposure and general exposure in buildings open to the public. In the latter case, in France for example, a threshold of twenty-five fibers per liter, subsequently reduced to five fibers per liter, has served as the benchmark for whether remediation of asbestos-containing buildings is required. Public authorities have long claimed that exposure below this threshold is safe.

In the United States in 1976, the Occupational Safety and Health Administration (OSHA) reduced the "permissible exposure limit" (PEL) for asbestos from five to two fibers per cubic centimeter (two thousand fibers per liter). A year later, France followed suit by creating a new "occupational exposure limit" (OEL) for asbestos at the same level. Note that this limit for workers is actually four hundred times higher than that set for the general public. Experts and policy-makers created these new regulatory policies with full knowledge that the new limit values would only protect workers against the risks of asbestosis but not cancer risks.[41]

Since the 1950s, scientists have argued that threshold limit values cannot be used to eliminate cancer risk, since just one errant cancer cell, possibly the result of exposure to just one molecule of a carcinogen, could lead to full-blown disease.[42] In this sense, the toxicological model of "the dose makes the poison" simply does not apply to cancer. What the thresholds produce, instead of risk-free management of exposure, is a political solution, one that establishes a trade-off between public health and economic growth.[43] In other words, because asbestos is a carcinogen, the use of limit values rests on a tacit agreement to place a certain number of peoples' health and lives at risk in exchange for the protection of industries that manufacture and use asbestos, as well as the benefits that asbestos use confers in several economic sectors. The tacit agreement is built on and reinforces the presumption that "safe" occupational exposures exist, even as those values maintain significant levels of risk for asbestos industry workers and legitimize the occupational exposures that will result decades later in a growing number of asbestos-related diseases.[44] As Ulrich Beck has put it, "acceptable [exposure] levels certainly fulfill the function of a *symbolic* detoxification" but without necessarily offering genuine protection.[45]

Limit values also contribute to the continued accretion of asbestos residues in the environment by ignoring how easily and how far the contaminant travels—out of factories and beyond construction sites. Moreover, the regulatory approach to controlling asbestos exposures overlooks many other groups that are unwittingly exposed to asbestos. The first group are so-called para-occupational exposures. These affect people living with or near asbestos workers—their families, relatives, and neighbors. It is easy to imagine a factory worker's spouse who inhales asbestos from the workers' blue coveralls while doing the family laundry. A case that gained widespread public notoriety in France in the 1990s involved a young butcher who died of mesothelioma at age twenty-eight. He had been exposed to asbestos as a child from playing with neighborhood children whose father was a worker in an Eternit factory in northern France.[46] Another asbestos victim group includes people living in places where asbestos is mined or processed—like Ambler, Pennsylvania—and where the fibers circulate at relatively high levels through the natural environment. Exceptionally, as in New Caledonia, Spain, or various counties in California, the source of asbestos may be a natural outcropping of the mineral linked to certain geological features of the soil. Because the source of these exposures is the natural environment rather than industry, regulatory bodies rarely take them sufficiently into account until or unless citizens mobilize to call political attention to the problem.[47]

Asbestos residues and exposures are organized by the rhythms of human labor and leisure, moving with miners and factory workers through social networks of human interaction. Such interlocking patterns of accretion are illustrated by the history of a small mineral grinding plant in Aulnay-sous-Bois, a town located in Seine-Saint-Denis, an industrial suburb of Paris. Named the Comptoir des minéraux et matières premières (CMMP), this plant processed asbestos from 1938 to 1975 and finally closed in 1991. At times, hazardous conditions at the plant became so severe that processing was interrupted by protests, letters, and petitions. Asbestos was actually moving out from the plant to contaminate the community—as evidenced by a nearby school and cemetery coated with white dust (see figure 3.4). In response, both local residents and elected officials expressed ongoing concern about the asbestos over several consecutive municipal administrations. Despite many injunctions issued by local authorities demanding that the situation be remedied, the firm did little to address environmental conditions inside

FIG. 3.4. Photograph of the Comptoir de minéraux et matières premières (CMMP) plant, which is next to a cemetery in Aulnay-sous-Bois. © Nicole et Gérard Voide, Collectif des riverains et des victimes du CMMP, 2005.

or outside the plant. When an individual living near the plant died of mesothelioma in 1997, residents, workers, and parents of school students organized their activism on a larger scale. Their protests against local and national authorities prompted an epidemiological study, conducted by the Institut de veille sanitaire, to collect data on this situation, including historical data.[48]

In the resulting 2007 study, epidemiologists documented a direct link between asbestos occurrence and community rates of pulmonary disease and cancer, estimating that exposure to asbestos may have affected several tens of thousands of people living in the vicinity of the plant. The study also offered a striking picture of plant conditions during its operating period. It included workers' testimonies that the atmosphere inside the plant was often extremely dusty and protective devices such as vacuum cleaners were seldom used. Empirical measurements of dust taken before the 1970s documented extremely high levels ranging from several hundred fibers per cubic centimeter to more than a thousand—far exceeding the limit values set into place a few years later. In addition, the study surveyed factory workers' medical records and found that at least twenty former employees at the small plant suffered from asbestosis, lung cancers, or mesothelioma.[49] Despite the public outcry that prompted this and several follow-up studies, few provisions have been made to inform residents who may have been exposed.[50] As a result, their illnesses are unlikely to be attributed to asbestos exposure.

It is impossible to accurately predict the future of this epidemic. The number of asbestos-related diseases is highly correlated with asbestos use, but their diagnoses and progression lag thirty to forty years behind exposure. While many journalists and analysts have stopped paying attention to the epidemic of asbestos-related diseases, asbestos exposures will in fact continue to impact millions of people for decades to come. In the UK, for example, asbestos-related disease was predicted to increase until about 2020, even though imports of asbestos have been declining since the late 1970s.[51] In France, which enacted a complete ban on asbestos use in 1997, we can expect a very slow decrease in the annual number of asbestos-related diseases beginning around 2030. But even if asbestos was banned worldwide tomorrow, global use and accretion of asbestos residues ensure that the epidemic of asbestos-related diseases will persist at least through the turn of the twenty-second century.

In asbestos, we see how a known hazard becomes invisible as it moves from economic domains of factories and warehouses to the domestic and community domains of household and neighborhood, blurring the boundary between exposure at home and at work. As a consequence, it slips through the gaps between occupational and environmental regulation, a commonality that asbestos shares with rare earths, our next case. The circuitry of asbestos accretion is easy to miss because it is time-lagged and spatially discrete. But its stunning human toll—mapped here through social networks of work, through disease occurrence and death certificates and litigation—raises a moral imperative to pay attention.

Rare Earths

Sometimes a geopolitical crisis introduces the public to the less familiar parts of the periodic table of chemical elements. This happened with radioactive elements in the 1940s, and it happened with rare earths in 2005. That year, China, the world's leading rare earths producer, began cutting export quotas for these elements by half and simultaneously instituted a new series of export taxes on those products.[52] Several western nations, whose economies depend on these critical materials, feared ever-rising prices and interruptions in supply.[53] As a *Le Monde* journalist put it at the time, "China has no oil but it has rare earths."[54] Its decision to show some economic muscle by limiting the global supply of these strategic minerals triggered a political crisis that, for a time, brought rare earths to the forefront of public awareness.[55]

Our final story in this chapter ties residue accretion to the global minerals markets and the commodity chains that spread rare earth minerals around the planet.[56] As residue, rare earths are interesting because they are risk multipliers, their globetrotting habits sustained by high-volume flows of toxic chemicals and heavy metals that are required for extraction, separation, purification, recombination, and recycling. The mountains of by-product waste generated by these processes fundamentally alter landscapes and threaten the survival of nearby communities. Following rare earth residues as they move from place to place shows that what is discounted by economists as "externalities" is undeniably experienced as the human costs of production and global trade. And the case of rare earths sheds light on often-ignored sides of the cleaner and sustainable technologies that environmentalists generally prize.

The term *rare earths* refers to a group of seventeen chemical elements, fifteen in the lanthanide series and two additional elements—scandium and yttrium—that share similar chemical properties. First identified by chemists in 1794 as "earths," an old reference for acid-soluble elements, they were considered "rare" at the time of their discovery because they were difficult to separate from ores.[57] Since then, geological investigations have shown that rare earths are widespread in the earth's crust and so in that sense are not rare at all. But they remain very expensive to produce and so are rare in a different sense, despite increased efficiencies in technologies used to extract and refine them.[58]

Not unlike asbestos, the growing fascination that scientists and engineers have shown for rare earth elements is related to their physical and chemical properties. Added in small quantities to other metals, they reinforce or enhance a metal's thermal and electrical conductivity, magnetism, or luminosity. These properties—used to engineer lighter, more resistant, and more conductive materials—give rare earths strategic value to militaries and key industrial sectors such as electronics and solar energy. These same qualities make rare earths economically valuable for use in hundreds of common and not-so-common consumer products—from televisions, computers, and mobile phones to electric vehicles and wind turbines.[59]

Global demand for rare earths rose sharply in the 1990s, leading to an increasing international flow of these materials. Before 1960, world production of rare earths was estimated at less than 10 kilotons per year; today annual production totals somewhere between 110 and 143 kilotons.[60] But "world production" has a specific geography (see figure 3.5). Through the 1980s, the United States dominated world production of rare earths, extracted from Mountain Pass, a single mine in California. In the 1990s, the political geography of rare earth extraction shifted as China ramped up production and drove down prices. Until 2015, around 85 percent of rare earth production came from Chinese mines and factories.[61] From there, rare earth oxides and nitrates are processed in a dozen other countries and sold in purified form to manufacturing facilities around the world. The purified rare earths are then combined with other metals, plastics, and glass to become the cell phones we carry in our pockets, the navigational systems built into our dashboards, or the LED bulbs that brighten our homes.

True world travelers, rare earths move from country to country along complex commodity chains. These chains are created by

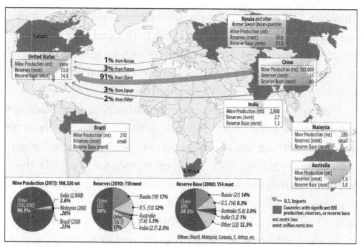

FIG. 3.5. Rare earth elements: World production, reserves, and U.S. imports. *Source:* Marc Humphries, *Rare Earth Elements: The Global Supply Chain,* U.S. Congressional Research Service, Report 7-5700, December 16, 2013, p. 12, based on data from U.S. Geological Survey, Mineral Commodity Summaries, 2008–2013.

an international division of production and trade that extends the reach of rare earth residues and their related toxicities. The health risks and environmental pollution associated with this industrial activity have not gone unnoticed but are not adequately documented. Several studies have drawn attention to the limited toxicological and epidemiological data that are available to assess the potential human health effects of these substances.[62] Moreover, the ongoing exposures and pollution levels are expected to remain significant on account of the scale and technological requirements for rare earths extraction and purification, as well as the massive waste their various forms and uses generate.[63] Reporting from "an artificial lake filled with a black, barely-liquid, toxic sludge" on the edges of the Baogang Steel and Rare Earth complex in Baotou, China, BBC journalist Tim Maughan confesses, "I'd seen some photos before I left for Inner Mongolia, but nothing prepared me for the sight. It's a truly alien environment, dystopian and horrifying. . . . It feels like hell on Earth."[64]

At open pit rare earth mines like the huge one at Baotou, production involves an unrestrained technological assault on the landscape. Industrial-scale hydrometallurgical processes and acid baths are used first to separate low concentrations of rare earth elements from mined rock and then to purify the minerals so that they can be sold on domestic and international markets. Up to 7 tons of ammonium sulfate and 1.5 tons of oxalic acid are needed to produce a single ton of rare earth oxides.[65] In 2017 alone, Chinese production facilities generated 105,000 metric tons of rare earth oxide.[66] The volume of wastes generated from this production is staggering.

The muddy, toxic by-products of separation and purification are pumped into tailings ponds like the one Maughan describes, which covers ten square kilometers. The sludge contains waste oxalic acid and sulfuric acid, a by-product of ammonium sulfate. Acid water seeps from such "ponds" and other waste storage sites, modifying the pH of the surrounding soil, increasing the oxygen consumed chemically, and silting up nearby rivers with sludge laced with lead, arsenic, and other heavy metals as well as (naturally occurring) radioactive thorium and uranium. The result is the massive contamination of hydrogeological systems that extend far beyond the mine site and tailing pond and the subsequent poisoning of entire regions. Mining communities have been afflicted with high rates of leukemia, genetic malformations, and other diseases, resulting in regional depopulation. Health impacts have become so severe that Chinese authorities have banned the use of water from the impacted areas.[67]

This is an example of residues thickening up, in the worst way imaginable. It is accretion on a massive scale, horrifyingly visible, yet few outsiders will ever witness it—an invisibility stemming from sheer geographical isolation rather than the microscopic nature of asbestos fibers or a subterranean plume of gasoline. But Baotou is only one link in the rare earths commodity chain. We take you next to La Rochelle, France, a different link, to illustrate another way that chemical residues pile up and slip away, invisibly.

La Rochelle is a coastal town in the southwest of France whose harbor opens onto the Atlantic Ocean. A far cry from the devastated landscape of Baotou, the city's commitment to sustainable development policies has given La Rochelle a reputation as an ecologically progressive "green" city.[68] At least, this is the image conveyed in the tourist marina located in the city center. A different image greets those who venture away from the center of town and wander down to the commercial and industrial port area. There, the image of La Rochelle as an experimental model of a sustainable city fades into working-class neighborhoods that surround a "high risk" industrial area concentrated with storage tanks for nitrate fertilizers and oil as well as an old and large rare earth production factory. Few people, including some residents of La Rochelle, are aware of this plant, even though it was once one of the world's most important rare earth production facilities.

Since its official creation in 1948 by the Rhône-Poulenc group, at the time the most important chemical and pharmaceutical company in France, the La Rochelle plant has been dedicated to the production of rare earths. By the early 1970s, the plant was supplying nearly half of the world's yttrium and europium markets and 80 percent of the lanthanum and praseodymium markets. Today the plant is controlled by the Belgian chemical multinational company Solvay. It specializes in the production of rare earths purified from the rare earth oxides imported after initial processing from China. The site also houses Solvay's industrial laboratories for research and the development of rare earth products. But despite major investments in green chemistry, the La Rochelle plant, like the Chinese facilities, has a waste problem.

For the first decades of operation (until 1974), the plant discharged its liquid and solid wastes directly into the sea without consideration of their effects on ecosystems. Especially because these so-called historical wastes are naturally radioactive, first Rhône-Poulenc and then Solvay have been asked by French regulatory statutes to manage the waste more responsibly.[69] In French regulatory language, rare earth wastes are distinguished by their "residues"—an official regulatory classification based on the legal status of the wastes' constituent elements. The most problematic are wastes containing radium, which is highly radioactive and emits radon.[70] Radium waste was removed from the plant and stored in a nuclear waste disposal facility in Normandy until 1990, when the facility reached capacity and was closed.[71] As a temporary solution, the French Atomic Energy Commission (CEA) agreed to store 5,090 tons of additional radium waste at the French Atomic Energy Cadarache site in Provence. The waste is still there.

La Rochelle also produced low-level radioactive waste that French regulatory authorities refer to as "banalized solid residues."[72] Considered less risky than radium waste, banalized solid residues were dumped in a landfill in the port area with annual authorization from the French public authorities. This dumping continued until 1993, when the city determined that the landfill had reached capacity and closed it. The actual amount of waste still contained in the municipal landfill is unknown but officially estimated by the French Nuclear Safety Authority at 61,000 tons. An unknown quantity of this same type of residue was mixed with backfill material and used to expand La Rochelle's industrial port, known as La Pallice. The La Rochelle plant itself contains another 8,400 additional tons of banalized solid residue, 21,750 tons of raw thorium hydroxides, and 16,400 tons of "suspended matter" whose contents are 25 percent rare earths. In total, more than 100,000 tons of chemical rare earth waste has transformed La Rochelle's industrial port, unbeknownst to most local residents.[73]

The company and regulatory authorities closely guard knowledge about the rare earth wastes from La Rochelle—including their specific locations, volumes, contents, and toxicities. Because regulations institutionally separate responsibility for managing hazardous waste from managing radioactive waste (a similar strategy used by regulatory officials in the asbestos case), only those managers and researchers in charge of radioactive waste have paid attention to the rare earth waste; to date, these actors have expressed relatively little concern with issues of chemical toxicity.[74] Basol, a national database maintained by the French Ministry for the Environment, indicates in closing remarks bereft of detail that the plant's activities have led to the migration of several chemicals and heavy metals into surrounding soils and groundwater.[75] These include ammonium salts, arsenic, mercury, nickel, and lead, among others. Measurements indicate that the contaminants are present in the environment at levels that exceed permitted standards. The limited knowledge available to the public is the result of pressure from anti-nuclear activists who mobilized around issues of radioactive waste, not the result of company action or efforts by regulatory authorities to systematically implement a management policy or anticipate problems of environmental risk or threats to public health. As such, knowledge about rare earth waste remains rare, even if the elements themselves are increasingly ubiquitous.

In China, France, and other countries as well, the accumulation of rare earth waste represents a mounting crisis. Yet disposal solutions remain out of reach. The limited exports of rare earths imposed by China in 2005 gave other countries new incentive to launch more sustainable disposal policies, and many began exploring the recycling of rare earths from waste. Today, though, recycling of rare earths occurs on a very limited scale. In 2011, the United Nations Environment Programme (UNEP) reported that less than 1 percent of industrial rare earths are recycled.[76] The recycling of end life products is often difficult due to the very low quantities most products contain—for example, a smartphone contains only fifty milligrams of neodymium and ten milligrams of praseodymium. In addition, rare earths are mixed with impurities in the final products, which makes their recycling difficult as well as polluting. To identify more efficient and less hazardous treatment processes the United States, France, and Japan have launched new research activities with financial support from companies such as Solvay, Umicore, and Mitsubishi as well as public authorities such as the European Union. Despite technical improvements, recycling remains too costly—and the costs of recycling too uncertain—to implement at scale.[77]

With rare earths, accretion follows global supply chains, and its patterns of accumulation and diffusion multiply with dizzying speed. Our brief excursions to Inner Mongolia and southwestern France have only scratched the surface. The colossal problem of rare earth waste, especially e-waste from the electronics and renewable energy industries, is mostly uncharted territory but poses serious risks and raises important questions of environmental justice.[78] The residues of e-waste are both thin, as found in the micrograms of rare earths inhabiting discarded electronics, and thick, in the massive ponds of toxic by-products from mining and processing. The long-term costs of their accretion remain unaccounted for in the global marketplace, even as seeking less toxic materials for electronics or recycling rare earths remain active industry goals.

Conclusion

Though not your typical travel guide, this chapter has offered a whirlwind tour with three residues through four cities on three continents. If the journey is the goal, then the goal of this journey has been to tease out common patterns of accumulation and diffusion in residues as distinct as gasoline, asbestos, and rare earth elements. The dissimilarities are undeniable, as the cases make clear. But there are also common threads that tie the three stories together and reveal larger patterns.

One commonality is their tendency to travel. Residues make circuitous, often globe-trotting journeys throughout their life cycles as chemical commodities on capitalist itineraries of production, consumption, and disposal. First mined and industrially processed for economic use and profit, residues follow engineered infrastructures and transoceanic trade routes. They move not only through worldwide supply chains but also through subterranean geologies and along sidewalks connecting work to home. Embedded in the products of our carbon-based material culture, for a time, residues become our cars, houses, cell phones, and light bulbs. When their utility ends, residues enter anew into waste streams that take them, eventually, to the lake of Baotou, the harbor of La Rochelle, or beneath your university building. Such processes knit together time and space, linking the accretive dimension of residues to the historical legacies described in the previous chapter.

Paying attention to accretion also spotlights the material transformations entailed in making stuff residual. There are many stages of materiality involved and at stake. As residues move, they may change from ore to computer components to electronic waste, for example, or from an asbestos ore, which in many cases is not harmful outside human bodies, to a concentration of fibers that becomes toxic inside bodies. The circulation of residues is determined in part by their own physical properties, whether organic or inorganic, liquid or solid, volatile or inert. The interactions of residues with the natural environment—in this chapter, weather, landscape, soil structure, and hydrology—also matter a great deal in understanding the social and ecological forces that keep residues on the move.

In different ways, for different reasons, and with different consequences, residues find their way to unexpected and sometimes unwelcome destinations. Yet we rarely anticipate the ways in which residues escape the systems constructed to contain and manage them. They scatter or pool here and there, often unseen but sometimes inconveniently surfacing. The uneven manner in which residues move around is an important feature of accretion. Whether as fugitives or dumped waste, residues often lurk beyond the realm of the visible, although their elusiveness, too, takes different forms and serves different political and economic ends.

Patterns of accretion spotlight the slipperiness of residues. And because of their slick mobility, we should not be surprised when residues show up unannounced—across the ocean, down the hill, or in your body. They can become invisible through transformations of land uses, their relocation to landfills, or the devices that regulate them. The stuff hidden out of sight (and out of care and thought) more often than not emits quiet signals that require detection and interpretation, like the chemical stench that pervades a Brown University building's basement or the clicking Geiger counter that alerts regulators (but not tourists) to La Rochelle's radioactive rare earths. Apprehending the accretion of residues stretches the capacities of our scientific and political systems to their very limits and beyond, as we will see in the next chapter.

4

Apprehension

In the 1980s, long summer road trips through the (West) German countryside were necessarily followed by time spent scrubbing the layer of smashed insects from the car windshield that had collected along the way. New drivers are probably unfamiliar with this ritual, which was not exclusive to Germany. In recent years, a series of scientific studies have documented the dramatic decline in insect populations in many parts of the world.[1] The most impressive one may be the "Krefelder Studie," published in 2017 by the journal *PLoS ONE*.[2] Since the early 1980s, the study reports, total flying insect biomass has decreased in Germany by more than 75 percent, and in summertime specifically, by more than 80 percent. The authors point to the growth of large-scale, intensive agriculture and the concomitant increase in insecticide usage as primary causes. The loss of biodiversity and biomass does not mean just a cleaner windshield, however. Insect-eating birds have also been decimated.[3] Bees and other insects needed for pollination or phytosanitary control are increasingly endangered.

How do we make sense of the news of vanishing insects?[4] There have been various responses. Some doubt that it has occurred, attribute it to other causes such as climate change,[5] or even depict the reports as junk science.[6] Others emphasize the development of solutions: replacement of insect pollinators by cheap female labor; the emergence of an insect industry for the agricultural sector; and the possibility, at least, of robotic drone pollinators.[7] Still others have pointed to the scientific studies as indications of an impending catastrophe, threatening agriculture on a large scale, jeopardizing global food supply, and ultimately serving as the harbinger of a major biodiversity loss, the "sixth extinction" in earth's history.[8] Even in the context of a daily human experience—in our example, roadway encounters with insects—perceptions of environmental change and interpretations of its meaning are divergent and conflicting. Understanding the seemingly enumerable consequences of this "slow disaster" and its probable cause—the expansion of industrial agriculture, including habitat loss and heavy reliance on insecticides—leads us into the tricky terrain of realizing, objectivizing, experiencing, fretting, and taking action.[9]

In doing so, we transition from the temporal and spatial dimensions of residues to consider how scientific and regulatory bodies, as well as local communities, encounter and make sense of them. As we have seen, our regulatory systems endlessly divide up both our assessment and management of potential problems of toxic residues, no matter how interrelated they may be. This hyper-segmentation often bears little relationship to the human experience of the world as it is. The division of the molecular world into discrete, manageable, molecular microworlds silences our firsthand knowledge of our surroundings. Simultaneously, it inhibits lines of inquiry that might integrate the complex mixtures of residual molecules already around and within each of us. If this fracturing prevents scientific and regulatory understanding of the world as it actually is, then what does it mean to apprehend residues?

To apprehend is to grasp that which has been elusive. In the case of a person, to apprehend is to hold; for a fugitive, it is to take into custody. Insights can be even more elusive than people, and to apprehend here is to take hold intellectually, to understand. Apprehension is also an affective and cognitive response to uncertainty: to be apprehensive is to fear. In approaching residues, it is these two aspects of apprehension—understanding and unease—that we find so intriguing and challenging to parse.[10]

In many cases, environmental contaminants are not detectable to humans through our senses. So we use science and technology to apprehend those residues that reside beyond our sensory apparatus.

But apprehension is also how we work to *make sense*, emotionally as well as intellectually, of our experience, including experiences that give rise to what Anthony Giddens has called "ontological insecurity."[11] Apprehension as sense-making is a collective, as much as an individual, activity: we reckon according to social, political, legal, and religious frameworks that attach meanings and judgments to our perceptions—even as these frameworks reflect different, sometimes irreconcilable, rationalities. News about environmental problems such as catastrophic insect loss often leaves one with a sense of foreboding that is, at the same time, pregnant with possibility: "Because I know this now, something must be done, but what?" When it comes to figuring out what to do with residues, our social technologies of sense-making rarely lead us to clear courses of action.

This chapter explores the multifaceted dimensions of apprehension through tools for detection, bureaucratic means of determination, and the very human experience of exposure. We enter into conversation with scholars who seek to understand the relationships of communities with the toxic environments they live in and the many ways they make sense of their experiences of these places, and even their own bodies, in ways that may be different from the senses of regulators, scientists, or activists.[12] In doing so we explore various, sometimes competing, ways in which different social constituencies deal with residues, construct them as (un)manageable problems—or not as problems at all—and (sometimes) imagine solutions. At the core of apprehension lurks the unruliness of residues but also the disconnections between different modes of approaching and understanding them.

Cases in this chapter illustrate three different and somewhat incongruous ways to apprehend residues. Sometimes apprehension is supported by the development or enhancement of new technical tools. As the first case of biomonitoring shows, the development of precise measurements of a chemical body burden and their application to broader population studies brings immediate attention—if not a sense of immediacy—to the connection between our bodies and our environments. Yet there are significant uncertainties in how to interpret biomonitoring information, which rarely informs policy about population-wide exposures. Instead, regulatory limits for "how much is too much" are specified through a highly social process of negotiation and interpretation, as illustrated in our second case by following deliberations of the European Union's Scientific Committee on Occupational Exposure Limits. At times, detection and determination combine in the experiences of individuals in direct and persistent contact with the chemicals of their trade. While biomonitoring provides post hoc data of personal exposure, often at a safe distance from the source of harm, agricultural workers exposed daily to pesticides have intimate contact with toxic residues. Their sensory experiences, as seen here through cases of glyphosate exposure, often stand in stark contrast to the regulatory representations of what it means to apprehend chemical environments.[13] As Donna Goldstein has observed, the invisibility of toxicity gives the state a key role in authorizing its existence.[14] More than our previous chapters, in apprehension, we encounter the close relationship between knowledge and power.

Monitoring Human Bodies

Human biomonitoring is an *extrasensory* way of apprehending contamination, one in which the repository for chemical residues is one's own body. Measurements of chemical body burdens obtained through population-scale biomonitoring have increased the knowledge of how thoroughly chemical contamination has permeated living organisms and augmented the anxiety around pollutants. The toxic environment is not (only) out there; it is *in us*.[15] Biomonitoring makes clear that residues have become an intimate problem. Whereas environmentalists of the 1960s and 1970s could point to the sensible effects of pollution, such as eye-burning smog and murky rivers, the contamination that biomonitoring turns up is rarely detectable to the human senses. Pristine-looking landscapes and healthy bodies, under chemical scrutiny, are shown to be harboring the molecular signatures of our accreted industrial legacy. The evolving detection limits and capabilities of biomonitoring continue to expand how and what we are able to apprehend.[16]

In the 1960s and 1970s, human biomonitoring emerged as a tool for assessing population-wide exposure to specific and targeted hazardous chemicals. Testing of human blood and urine for signs of chemical exposure has since become the "gold standard" of environmental public health, leading to ongoing large-scale studies.[17] Here we will focus on developments in the United States, though biomonitoring has also been conducted by national and international agencies elsewhere, especially in Europe. In the 1960s, population-based surveys of lead helped usher in stronger U.S. environmental protection, and subsequent biomonitoring data showed that this regulation was effective in decreasing exposures and improving public health. However, as biomonitoring has become a more diverse tool (testing for the presence of hundreds of chemicals simultaneously) the chemical industry has resisted giving weight to recent biomonitoring surveys that document the presence of numerous toxic chemicals at doses so low that their health effects—individual, let alone synergistic—are often unknown. Biomonitoring is rarely a direct input to policy. In general, it is not the tool that regulators use in formulating limits and rules, though it can be used to show the fallibility of those determinations. In addition, the individual nature of biomonitoring data can lead to a privatized conception of contamination, rather than framing exposure in terms of public health, leaving regulation in the hands of the consumer—and, consequently, industry.

The principal (and oldest) biomonitoring techniques use analytical chemistry to detect the presence of specific chemicals in human bodily fluids and tissues, especially urine and blood.[18] Sometimes, not only is the original compound sought, but also its so-called metabolites. The breakdown of chemicals by the body, especially through detoxification enzymes in the liver, produces these metabolites. These by-products can be more reliable signatures of exposure than the presence of the compound itself—and in many cases, these metabolites are themselves toxic in ways that the original chemical may not have been. Most population-wide screens rely on high-precision analytical instruments to measure the levels of chemicals or their metabolic residues. In fact, nearly all official reports and reviews point to post-1960 improvements in laboratory technology as driving the expansion of human biomonitoring. The equipment used in contemporary biomonitoring surveys conducted by the U.S. Centers for Disease Control and Prevention (CDC), for example, utilize a suite of cutting-edge instruments and novel, validated methods and techniques to enable residues to be detected at ever-lower quantities.[19] As Ken Sexton, Larry Needham, and James Pirkle have put it, "Specialists can now detect extremely low levels—parts-per-billion, parts-per-trillion, even parts-per-quadrillion—of multiple markers using a relatively small sample, say, 10 milliliters or less."[20]

While these new technologies have increased the breadth and precision of biomonitoring, the testing of human body products and tissues emerged in the nineteenth century as a means for diagnosis and drug calibration in patients—and to identify poisons in dead bodies for medical forensics.[21] In the first half of the twentieth century in the United States, industrial plants began to monitor workers' bodies via blood or urine tests to prevent poisoning from exposure to toxic substances such as lead and benzene.[22] Such oversight did not extend beyond the walls of the plant, and the physicians and toxicologists charged with worker safety were employed by the companies. In these respects, "industrial hygiene" remained disconnected from municipal public health efforts, which largely focused on sanitation and water safety.[23]

Concerns about lead poisoning prompted the first large-scale use of biomonitoring outside of industrial plants. By the 1950s, some pediatricians were reporting on a worrisome increase in lead poisoning of children, often from paint in poorly maintained apartment buildings.[24] Paint companies began reformulating their products to have lower lead levels and labeling cans, but paint was only one source of exposure. Public health agencies in the United States employed the same blood assay used to monitor industry workers to conduct population-wide surveys. Blood samples from "The Three City Survey (1961–1963)" showed that individuals living in urban

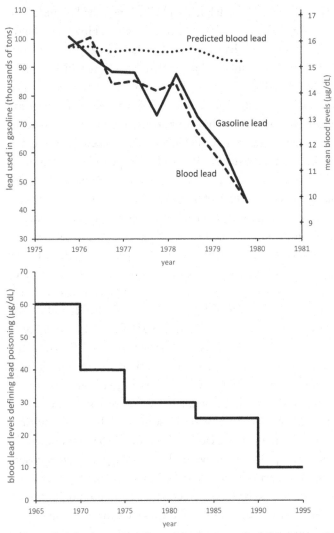

FIG. 4.1. Graphs showing, on the top, the decrease in leaded gasoline and in average human lead exposure (as measured in micrograms of lead per deciliter of a person's blood) in the U.S. population after 1976; and, on the bottom, the changing definition of lead poisoning (again, determined by micrograms of lead per deciliter of blood) from 1965 to 1995. When leaded gasoline began to be phased out in the 1970s, experts did not predict a large effect on lead exposure in the U.S. population. However, blood lead levels fell significantly, correlating strongly with the amount of leaded gasoline being marketed. During those same years, continuing studies of lead toxicity resulted in ever-lower concentrations of blood lead levels qualifying as lead poisoning. Figures and legend adapted from Ken Sexton, Larry L. Needham, and James L. Pirkle, "Human Biomonitoring of Environmental Chemicals," *American Scientist* 92 (2004): 38–45, 44.

areas (Los Angeles, Philadelphia, and Cincinnati) had higher blood lead concentrations than residents of rural areas. However, even in rural areas, individuals had some lead in their blood.[25] Geochemist Clair Patterson presented evidence that environmental lead, even in remote areas, derived from industrialization, not natural sources (as prominent industrial toxicologist Robert Kehoe had argued). Patterson extended his analysis to human blood levels, concluding that "the average resident of the United States is being subjected to severe chronic lead insult."[26] The lead story provides an especially vivid illustration of how the development of occupational health standards in the first half of the twentieth century provided the framework for environmental health studies in the second half.[27]

Biomonitoring was then added to preexisting public health surveys. Since the early 1960s, the CDC (part of the U.S. Public Health Service) has surveyed trends in health and nutrition status across the entire U.S. population.[28] Blood lead levels were included in the CDC's nine thousand-person National Health and Nutrition Examination Survey (NHANES II) from 1976 to 1980, as well as in its subsequent surveys. The data allowed for comparisons over a twenty-year period, documenting that the introduction of unleaded gasoline (for newer cars with a catalytic converter) in conjunction with EPA-mandated lower levels in leaded gas had resulted in a 70 percent decrease in overall blood lead levels.[29] Regulation—in this case, just lowering the amount of lead in gasoline—had proven effective in dramatically reducing exposure and potential harm.[30] After the U.S. Environmental Protection Agency completely banned tetraethyl lead in automobile fuel in 1982, there was a further drop in the average concentration of lead in blood from sampled Americans (see figure 4.1).[31] The experience with lead, which illustrated that a public health survey could document widespread exposure to an industrial toxic while simultaneously tracking the impacts of regulation, prompted further use of population-based testing for environmental health.[32]

Other U.S. government surveys focused on everyday exposure to synthetic chemicals, particularly pesticides. In 1967, the U.S. Public Health Service established a National Human Adipose

Tissue Survey (NHATS) that tracked changes in pesticide residues in the bodies of U.S. residents. (The most commonly used pesticides in the postwar period concentrate in fatty tissue, requiring a different sampling strategy than blood or urine analysis.) In 1970, this program was transferred to the newly founded EPA.[33] By 1992, twenty chemicals had been measured in all fourteen thousand people sampled and more than one hundred other chemicals in a smaller number of individuals.[34] This study, based on autopsy and surgical specimen tissues from major metropolitan areas of the United States, documented widespread pesticide exposure. The EPA's ban of PCBs in 1979, for instance, resulted in a dramatic decline in tissue concentration of these chemicals, and similar decreases followed restrictions on usage of DDT and dieldrin.[35] However, these contaminants did not completely disappear. Subsequent CDC NHANES surveys have tested for—and detected—biomarkers (the chemical itself or a known metabolite) for more and more toxic chemicals.[36] In the 2003 NHANES III survey, amounts of two pesticides actually exceeded the officially permitted dose level for chronic exposure, especially in some children.[37] The Pesticide Action Network of North America drew on these numbers to call for stricter regulations.[38]

Some environmental nongovernmental organizations (NGOs) have carried out their own human biomonitoring studies in order to demonstrate the ubiquity of exposure to toxic chemicals both through the environment and in everyday consumer products. In 2003, the Environmental Working Group (EWG) published *Body Burden: The Pollution in People*. Two physician-researchers at Mt. Sinai School of Medicine, Philip Landrigan and Michael McCalley, worked closely with the nonprofit research organization in designing their test of blood and urine samples from nine adult individuals for the presence of 210 different chemicals. By contrast to the CDC NHANES, the EWG study used a much smaller sample of people but looked for a greater variety of chemicals. Their study subjects contained detectable levels of "76 chemicals linked to cancer in humans or animals (average of 53), 94 chemicals that are toxic to the brain and nervous system (average of 62), 86 chemicals that interfere with the hormone system (average of 58)," and dozens of other known toxic substances and reproductive toxicants.[39] EWG emphasized the scientific legitimacy of its study, with results first appearing in the peer-reviewed journal *Public Health Reports*.[40]

Biomonitoring results do not have to be scientifically original to be politically useful. In 2004, the World Wildlife Fund (WWF) released results from testing thirty-nine members of the European Parliament for the presence of 101 different chemicals. Every single member was contaminated with at least one chemical from each of five groups: organochlorine pesticides including DDT, PCBs, brominated flame retardants, phthalates, and perfluorinated compounds (such as PFOS).[41] An additional survey of fourteen environment and health ministers from thirteen E.U. countries yielded similar results.[42] The WWF's targeting of European lawmakers was aimed at increasing public support for the European Union's proposed chemical law—Registration, Evaluation, Authorisation and Restriction of Chemicals—which came into force in 2007.[43] In these ways, biomonitoring has been leveraged as a tool for highlighting the legacy of our industrial past and the ways in which we all share in its chemical aftermath.

Even when human biomonitoring data are abundant and precise, interpreting them is fraught with problems. As one NRC (National Research Council) report in the United States puts it, "The ability to generate new biomonitoring data often exceeds the ability to evaluate whether and how a chemical measured in an individual or population may cause a health risk or to evaluate its sources and pathways for exposure."[44] For only a few contaminants, such as lead and mercury, does sufficient evidence from human or animal studies exist to allow for credible interpretations of health risk from very low-level exposure. For most of the chemicals under question, "there are no epidemiological data on the relationship between the biomarker and the effect."[45] Detecting a contaminant in someone's blood or urine does not indicate how it got there—through what route or kind of exposure (ingestion, inhalation, or dermal absorption). Scientific apprehension does not necessarily translate into culpability or prediction.

Perhaps unsurprisingly, industry has seized on the uncertainties that inhere in interpreting most biomonitoring results. As expressed by one spokesperson, "Industry sees a movement toward collecting a lot of biomonitoring data prematurely, before we know what to do with it."[46] The American Chemistry Council, a trade group, emphasizes that biomonitoring, while useful, "does *not* provide information about (1) where the exposure came from (2) how long a substance has been in the body or (3) what effect, if any, that substance may have on the body."[47] Moreover, there exists a bureaucratic gap between government surveillance of populations for contamination and oversight of industrial sources of these same residues. In the United States, the government agency that conducts the most thorough biomonitoring, the CDC, is not involved with regulating the production or environmental exposures from chemicals (those responsibilities fall to the EPA, Occupational Safety and Health Administration, and Food and Drug Administration). Most media coverage of the chemical body burden creates public fear without a clear path to action.

That said, some environmental justice groups have been able to employ human biomonitoring to demand state action, especially where there are strong environmental laws. In the case of one agricultural community in California, biomonitoring results were used successfully to stop the aerial spraying of the pesticide chlorpyrifos. In this case, the blood levels of twelve residents exceeded EPA's "acceptable" adult exposure limit, providing strong grounds for immediate action. However, as Raoul Liévanos, Jonathan London, and Julie Sze note in recounting this incident, progress on environmental justice using such methods has been uneven and relies on community-based participation in biomonitoring as well as responsive government agencies.[48] Sara Shostak's study of a public housing complex in Midway Valley, California sheds further light on the challenges of relying on biomarkers to achieve legal action. Residents there were concerned that chemicals in the soil had contributed to endemic health problems, and chromosomal analyses of individuals there showed DNA damage of the kind associated with carcinogenesis. However, chemical analysis of soil samples did not substantiate their claims of high exposure, even as other pathways for exposure remained unexamined. Results from human biomonitoring do not necessarily have legal standing in regulatory frameworks geared toward direct measures of environmental contamination.[49]

In most cases, human biomonitoring turns up residues of pesticides and other toxicants that are detectable but that fall below government-set limits, in a "grey zone of chemical exposure" exhibiting levels that are neither alarming nor "unequivocally 'safe.'"[50] In this case, it is unclear what people can or should do with this information. There is wide variation in whether individuals tested receive information about their personal results, information about group-level data, or no information at all. Researchers doing clinical studies do not always report results to individuals unless exposures can be connected to known health consequences. Community-based groups that engage in biomonitoring demonstrate a stronger "right-to-know" commitment, and one group reports that informing individuals of their chemical body burden, if communicated appropriately, need not result in excessive anxiety.[51] This was confirmed by a qualitative study of responses to biomonitoring information concerning exposure to household chemicals among a group of women in Cape Cod, Massachusetts. While responses varied, few women who learned of their exposures to household and environmental chemicals showed alarm, though many were frustrated at the lack of available scientific knowledge about the health consequences of their particular exposure levels.[52] The authors suggest that just as individuals have different disease experiences, so too do they have different "exposure experiences." Consumer norms about personal hygiene and landscaping may in fact lead to habituation to exposure, especially from everyday chemicals.[53]

Nicholas Shapiro, Nasser Zakariya, and Jody Roberts raise the issue of whether demands for more environmental testing valorize quantification for its own sake rather than asking broader questions about why so many toxic chemicals are being produced in the first place.[54] This is an important issue for individual biomonitoring, especially as services become commodified. In 2006, a person could have their blood tested for a battery of 320 chemicals at a cost of $15,000.[55] More recently, the Silent Spring Institute has begun to sell a "Detox Me Action Kit" for $399 that tests an individual's urine for ten common chemicals found in consumer products, all suspected endocrine disruptors.[56] The institute encourages purchasers to download a "Detox Me Mobile App" that instructs them how to modify their lifestyle to reduce exposures and then to purchase a follow-up urine test to look for improved numbers. This individualistic approach to controlling exposure colludes with a broader neoliberal orientation in environmental politics that sociologist Andrew Szasz has criticized as "shopping our way to safety."[57] Consumer biomonitoring enables an appreciation for one's own burden of chemical legacy but inhibits a broader apprehension of where this legacy comes from and how publics might take collective action to decrease exposure generally.

Setting Occupational Exposure Limits

As we have just seen, comparing human biomonitoring results with thresholds set by state agencies may be used as a way to alleviate the worries associated with the presence of contaminants inside human bodies, to determine levels of contaminants that are safe—or, at least, "acceptable"—and, in doing so, to normalize or naturalize exposure. The apprehension of chemicals involves not only scientific instruments, as those used in biomonitoring, but also regulatory tools—like the establishment of so-called occupational exposure limits—that allow science to be incorporated into public policies. The purpose of such regulatory tools is to enable political management based on scientific knowledge. This process of domestication renders chemicals and chemical products administrable and manageable. Like other regulatory tools, limit values contribute to the shaping of the apprehension of residues. They officially arbitrate between various understandings of the presence of toxic chemicals in the environment and thus constitute either a resource or stumbling block for those engaged in toxic issues. While limit values exercise their political power through their foundation in science, the process of constructing—or determining—these values is anything but strictly scientific.

As mentioned earlier, regulatory tools create demarcations: between substances, their effects (hazardous or safe), their legal status (drug or food), and their use (at work or in the environment). These distinctions themselves are the residual legacies of our use and understandings of the molecular world. Thus a single compound investigated within the framework of agricultural production is studied, measured, and understood differently than when the same molecule appears in a closed workspace or is involved in a pharmaceutical application. These boundaries are actively reinforced as the tools, techniques, and knowledge contained within them develop an internal momentum and logic. With such reinforcement, our ability to apprehend becomes clearer in cases where, for instance, thresholds or limit values are used to limit individuals' exposure. Since the 1950s, for instance, national laws have set population-wide limits for exposure to ionizing radiation, but those limits are different (in the United States, an order of magnitude lower and thus more protective) than the limits set for occupational exposure.[58] Clarity in one economic domain can magnify uncertainty in others.

The first limit value was calculated in Germany at the end of the nineteenth century for carbon monoxide; it was also there that the first lists of limit values were published, in 1938.[59] Systematic work for establishing limit values in occupational health developed significantly in the United States during the first half of the twentieth century, especially in the industrial hygiene department of Harvard University.[60] Starting in 1946, the American Conference of Governmental Industrial Hygienists (ACGIH) published a list of what, after 1956, they called threshold limit values (TLVs), which quickly became the reference in many countries. (The ACGIH later trademarked TLV, which may account for the use of other terminology such as occupational exposure limits by other organizations.) During the 1960s and 1970s, limit values were introduced into different regulatory sectors at the national level in the United States, Germany, and other industrialized countries. In France, though, their introduction into government regulation came much later; it was not until the mid-1980s that the Ministry of Labor initiated an investigation of the subject and its potential applications. For the French Ministry of Labor, the objective was to produce a uniquely French set of limit values even while the majority of companies used values published by the Unites States' ACGIH. Under the impetus of the European Union, the French government began specifying limit values in its official regulation in the 1990s to 2000s.[61] Today, even though their use is combined with other regulations such as the obligation to substitute known carcinogens with less hazardous substances, limit values occupy an increasingly important place in French (and more broadly, European) regulation of employees' exposure to dangerous chemical products. As such, they have played an important role in legitimizing workers' exposures to hazardous chemicals, as well as claims by companies to be meeting those standards.

Social science research on limit values has developed a serious and sustained critique of the heavy reliance by the state on these tools for at least four reasons. First, there is not enough attention to how the values are established. Researchers have cautioned about and, at times, strongly denounced a system that they see as overly reliant on expert advice from industrial stakeholders. The obfuscation of process in favor of reliable output generates serious concerns about the validity of the established standards.[62] Second, limit values are often determined without attention to how the standards may be imposed or implemented. Rule-making, implementation, and enforcement all entail potentially dangerous and unpredictable lag times. Third, the hyperfocus on the generation and evaluation of a scientific value and its reinsertion back into complex social, economic, and political contexts masks other worker and environmental concerns that can become lost in the extraction process. This, in turn, can render limit values inapplicable to lived realities of workers, who are often exposed to many toxic substances (whose interactions may be unknown) under conditions quite different from those of laboratory experiments used to determine the limit values. Finally, the contested nature of the studies required to develop limit values often makes the entire process bureaucratically inefficient—often taking many years and untold financial resources to generate regulatory action on just one molecule.[63] In short, the particularity of the values and the long process required to obtain them means they cannot be extended readily even to chemical classes, let alone to the chemical industry as a whole.[64]

One way to understand how occupational safety standards are set is to observe the ways experts actually establish OELs. The Scientific Committee on Occupational Exposure Limits for Chemical Agents is one of the oldest expert groups set up by the European Commission. It was officially created in 1995 after its informal existence between 1990 and 1995 as a "scientific expert group." Such expert groups were commonly established at that time in European governance before the bovine spongiform encephalopathy (BSE) crisis prompted the creation of more structured agencies (such as the European Food Safety Authority [EFSA] or, more recently, the European Chemicals Agency [ECHA]).[65] During this earlier period, expert groups advised the Commission by providing scientific expertise, but they also gathered information from stakeholders. At the time, conflicts of interest did not receive the same attention as today, and experts were not required to declare their links to industry. Moreover, the fact that SCOEL produced standards for occupational health (rather than consumer safety) helped keep its activities out of the public eye. As the BSE crisis showed, once issues such as food safety or environmental contamination become public and political, stakeholders can no longer easily resolve the crisis.[66]

In 2014, a new European Commission decision was published that changed most of the rules organizing the recruitment of experts and deeply affected the organization of SCOEL in establishing OELs.[67] Observations of this committee during the last meetings before the implementation of this new regulation (in 2014 and 2015) revealed that technical determinations are entangled in real time with bureaucratic exigencies and commercial interests.[68] Decisions have to be made under time pressure and are not necessarily aligned with expert assessment. For these reasons, following the actual decision-making process sheds new light on the apprehension of toxic chemicals inside regulatory bodies rather than human ones.

SCOEL is a group based directly on the work of experts. There is no administrative staff working with the experts as is the case in agencies such as ANSES (the Agency for Food, Environmental, and Occupational Health and Safety) in France or IARC (the International Agency for Research on Cancer) at the World Health Organization (WHO). The only person helping the SCOEL chairman is the scientific secretary, appointed by the European Commission. Each expert working on a report must find help on his or her own. This point largely explains the inequality between experts: the most active are those who are supported with research assistance from their home research, regulatory, or academic institution. This explains why the most active experts come from Germany, the Netherlands, and the Scandinavian countries, which host major occupational health research departments.

When reviewing the case for an existing value, SCOEL has to determine whether new toxicological, epidemiological, or experimental studies in humans result in a threshold change. While in general, OELs have tended to decrease over time,[69] during the period of 2014 to 2015, discussions took place within the SCOEL regarding two limit values that had been targeted by industrial actors as too low. The first case involved aniline, an organic compound used to manufacture precursors to polyurethane and other industrial chemicals. SCOEL first published a report on this substance in 2010. It proposed an OEL of 0.5 parts per million aniline in a cubic millimeter (ppm/mm³). The main data backing the choice of this value came from a study conducted fifty years earlier, in 1961. Later, industry financed a new human experimental study in Germany involving the laboratory of the SCOEL chairman.[70] One of the conclusions of the study, based on experimental exposures of volunteers in human whole-body exposure chambers, was that an OEL of 2 ppm/mm³ was sufficiently protective of workers.[71] In the March 2014 meeting, the chairman presented the study, noting that, "We have a dispute with industry. They submitted a research report which says that, based on human studies, SCOEL's OEL is too low. I think the OEL should be at 2 ppm but we have to wait until this study is published."[72]

SCOEL members discussed several points during the next meeting and decided to increase the OEL for aniline from 0.5 to 2 ppm/mm³. Some experts insisted that this new study did not address the carcinogenic risk and expressed criticism regarding the design of the new study. For example, some wondered, why was the pilot study (with two male and two female volunteers) based on eight hours of experimentation while the final study (ten male and nine female volunteers) was based on an exposure of just six hours? With regard to this question, the chairman argued that this study was scientifically satisfactory: "They did what they could."[73] Another question concerned whether the volunteers were at rest or exercising during the experiment. Some of the inconsistencies were clarified in the written version of the report. Despite numerous discussions about changes in the arguments of the report, the new limit value of 2 ppm was never challenged or discussed.

The second chemical for which SCOEL discussed increasing the limit value was formaldehyde. In this case, the SCOEL report establishing an OEL was adopted in 2008 and specified the OEL (average of eight hours exposure) at 0.2 ppm and a short-term exposure limit (STEL) at 0.4 ppm. The committee's decision was based partially on a human experimental study with ten male and eleven female volunteers funded by industry and published in 2008.[74] For industry, this OEL was also too low. So they funded a new human experimental study in which scientists from the same laboratory discussed the OEL set by SCOEL. In their conclusion, they wrote, "We disagree with SCOEL's interpretation of results obtained in our previous study."[75] The method of the study was explicitly designed to produce a new OEL.[76]

In the March 2014 meeting, one of the scientists in charge of the industry-funded study was invited to give a presentation to SCOEL. It is not common for scientists who are not expert members to attend the meetings, and some SCOEL members seemed reluctant to listen to this presentation. The chairman explained just before the presentation that industry had requested the presentation. "However, we have already considered this study in our report. But if there is a study that can make us change our minds, it is that one."[77] Thus this visit was essentially an exceptional opportunity for industry to make its case in person.

SCOEL also discussed formaldehyde during three subsequent meetings. There were numerous discussions about the experimental study and a recent meta-analysis of epidemiological studies was also a subject of interest.[78] According to the expert presenting this new meta-analysis, the study showed no clear relation between low-dose exposure to formaldehyde and cancer. Here is how an industry expert proposed a new OEL during the December 2014 meeting, based on animal experimental studies:

EXPERT 1: In this study, we observe a non-linear exposure-response relationship. So we have to review the existing cohort studies. [. . .] As we find no dose-effect relationship in epidemiological studies, we have to rely on animal studies. [. . .] Our current OEL is not scientifically based. We could put 0.3 instead of 0.2. [. . .] It could be based on a NOAEL [No Observable Adverse Effect Level] of 1. We would apply a factor of 3 to obtain 0.3. With this value, we will have a complete protection against cancer. Even with 0.5 or 0.7, we would have no irritant effect. I would suggest 0.6 as STEL [short-term exposure limit].[79]

After this proposal, most experts agreed to the new level of the OEL. The decision was marked by informal assent and a sense of time pressure:

CHAIRMAN: We have to go faster. Next meeting will be our last meeting. 0.3 is an OEL. And STEL is 0.6. Are we okay?
EXPERT 2: I agree on 0.3 and 0.6. I think the 0.6 could cause problems later on.
EXPERT 3: OK for the values but the text needs to be completely rewritten.
EXPERT 4: OK with 0.3 and 0.6. But the text is unclear about the effects that we seek to avoid.
CHAIRMAN: The text will be polished.

EXPERT 5: On genotoxicity. This product is genotoxic but not at the levels that we are analyzing. But at higher levels, it is.

EXPERT 6: We need to explain why we do not put an interpersonal [security] factor.

CHAIRMAN: This point has already been discussed. We will make clear this point. [The scientific secretary] will present a polished release of the text.[80]

As the transcript reveals, experts' arguments over the report text are remarkably disconnected from setting the value. In essence, writing the report and establishing the OEL appear to be distinct activities, though the committee is responsible for both. More importantly, once an agreement is reached on the level of the OEL, a scientific argument is developed to justify it. In the March 2015 meeting, the chairman tried again to adopt a new recommendation on formaldehyde, but there was still some opposition. The new OEL was finally adopted in 2016 with 0.3 ppm as OEL and 0.6 as STEL.[81] Apprehension of exposure and danger, as determined through scientific studies, does not dictate regulatory decision-making, and reaching consensus in this scientific committee seems less about certainty and deliberation than efficient use of time and available knowledge.

The social dimensions of setting OELs go beyond the character of committee meetings to the way in which commercial interests are embedded in the process. The debate between experts is framed by scientific publications, which in many cases report on research funded by industry. In the cases analyzed above, the experts were persuaded to consider a less protective limit by viewing it as in better agreement with new data. These technical discussions are very far from an analysis of the actual situation of workers exposed to chemicals. In this respect, relying on an expert group such as SCOEL operates to create distance between bodies of workers, industrial plants, and pollutants (dust or chemicals) on the one hand and on the other hand, regulatory science, expert discussion about toxicological studies, and related debates about safety factors. Maintaining this distance within the context of expert deliberation about the scientific dimension of OELs is one way that regulatory bodies build legitimacy for the industrial use of chemicals. Despite the chasm, determinative practices imbue regulatory tools such as OELs with the veneer of scientific legitimation that itself is primarily the result of social compromises between public health and economic growth.

Experiencing Glyphosate

Many regulatory tools transform toxic residues into allegedly well-managed problems by creating an apparent distance between their biological effects and affected populations. Users are assumed to comply with whatever guidelines keep exposures below a limit value, and contaminants are assumed to remain contained in the context of use. Yet as we have just seen, such tools rarely incorporate the direct experience of individuals exposed to toxic substances, especially those who consider themselves victims of toxic legacies and accreted residues. Examining the uneven experiences of these communities allows us to watch chemical regulation unfold from the other side, where the unruliness of chemical encounters is sharpened by the actions of community activists and environmental NGOs.

N-(phosphonomethyl)glycine, more commonly known as glyphosate, is a versatile molecule. Synthetized in 1950, it was first sold as a chelator. Chelators possess the interesting (and sometimes useful) property of binding very tightly to metals. This binding creates amazingly stable molecules. These new molecules can then transport the stabilized metals to new places in the environment or in the human body. In the early 1970s, researchers at Monsanto found that glyphosate was a uniquely powerful herbicide, owing to its ability to inhibit protein synthesis crucial to growth in all plants and some microbes. The resulting product, known commercially as Roundup, has since been marketed as a cheap, safe, and efficient broad-spectrum herbicide and crop desiccant.[82] Today, glyphosate is the most widely used herbicide in the world. Since the Monsanto herbicide patent expired in 2000, glyphosate has become an integral component of many products developed by other companies.[83]

Glyphosate continues to be the cornerstone of Monsanto's product portfolio.[84] The generation, production, and distribution of the company's genetically engineered crops, known as Roundup Ready, have only further increased the sale and use of the herbicide. In 2014, the company sold more than five billion dollars of the herbicide with projections that this amount could double in the coming decade.[85] Globally, glyphosate use has risen almost fifteenfold since the introduction of these genetically engineered seeds.[86] The molecule and its metabolite AMPA (aminomethyl-phosphonic acid) have become globally ubiquitous environmental and human contaminants. For instance, in 2014, the U.S. Geological Survey announced that it had detected glyphosate metabolites in 75 percent of the rainwater and air samples it analyzed in Mississippi in 2007.[87] In France, glyphosates are the most commonly found contaminants in rivers.[88] Each biomonitoring investigation featuring glyphosate has registered it in human bodies.[89]

Just as the molecule has become globally pervasive, so have protests about its use. In part, this is due to glyphosate's major use in conjunction with genetically modified crops. But in addition, glyphosate has been associated with a vast array of detrimental effects on humans, farm animals, and the environment.[90] We will follow how protests against the widespread use of glyphosate have emerged in Europe, Argentina, and Sri Lanka. These efforts point to commonalities but also differences, not only in the regulatory systems and political responsiveness, but in the actual experiences those exposed to the substance have reported. And while there have been interactions among these sites and activists, there is also an asymmetry between Europe on the one hand and Argentina and Sri Lanka on the other. European activists have followed and mobilized around the toxic experiences of workers and communities in the Global South to strengthen regulation in Europe, but those changes do not necessarily relieve the situation of the suffering agricultural workers, who often have to contend with underfunded health care systems and inadequate regulatory oversight in their own countries. In other words, the internationalization of concern has not translated into a uniform or universal regulatory response.

On June 15, 2017, a petition against the use of glyphosate, initiated by a wide-ranging coalition of thirty-seven NGOs across Europe, had accumulated one million signatures from E.U. citizens.[91] This petition supported a "European Citizens' Initiative"—a procedure through which citizens may officially question the E.U. executive body, the Commission—demanding the ban of glyphosate herbicides in the European Union. The petition signaled the influence of antiglyphosate activism that had developed across Europe over the two previous years opposing the relicensing of the herbicide. Through the petition, NGOs—as well as committed investigative journalists—not only jeopardized the E.U. Commission's plans to quietly reauthorize glyphosate but also succeeded in documenting, chronicling, and analyzing the fierce political battles over glyphosate that took place between 2014 and 2017, which provide an eye-opening view into E.U. policy-making and regulatory practices.[92] The case of glyphosate shows that apprehension goes beyond the activities of scientific detection and regulatory determinations to include bodily experiences of exposure-related illness as well as political activism.

Over two years, advocates of a glyphosate ban in the European Union built support from a larger network of actors pursuing related but different agendas. In January 2014, the E.U. Commission launched the reauthorization process by submitting a report prepared by the German Federal Institute for Risk Assessment, which had been entrusted by the European Food Safety Agency (EFSA) with the assessment of glyphosate's safety.[93] The BfR recommended glyphosate's reapproval while refusing to disclose the two key chronic toxicity studies provided by industry that undergirded its decision, arguing that they contained commercially sensitive information.[94] NGOs strongly criticized the German assessment for failing to take into account other, peer-reviewed studies showing toxic effects.[95] The exposure of hidden EFSA practices such as "copy and paste" text substitutions from industry files into its official assessments and the acknowledgment by the EFSA director that these practices were normal strengthened resistance to the recommendation before the Commission.[96]

The challenge to the herbicide's reauthorization gained steam when the IARC (a unit of the UN's World Health Organization) classified glyphosate as "probably carcinogenic to humans" in March 2015.[97] In response, Monsanto organized an unprecedented worldwide campaign to undermine the IARC's legitimacy.[98] Monsanto's efforts partially backfired when the release of internal documents revealed their concerted efforts to discredit a long-standing U.N. agency. The disclosures reinforced the environmental movement's critique of corporate influence over regulation.[99] As public concern grew, images of Argentine children and workers afflicted with cancers and other abnormalities from exposure to Roundup in soybean agriculture found their way into documentaries and photo exhibits.[100] The images inverted the logic of biomonitoring in demonstrating glyphosate toxicity by making its victims and their suffering visible. Activist publicity in the European Union broadened the movement's support beyond antitoxics and environmental health NGOs, attracting attention from groups focused on human rights and trade policy and even from the European Parliament.[101]

Despite mounting opposition, the Commission narrowly approved the reauthorization of glyphosate in 2017. This controversial decision highlighted several related ongoing political issues. These included a much-sought-after reform of pesticides' assessment procedures and, more generally, of hazardous chemicals. It also made evident the lack of transparency of the E.U. Commission and the lack of E.U. institutions' independence from corporate actors. The decision also drew attention to the lack of democracy within the E.U. decision-making process.[102] Yet as we saw in the case of asbestos-laden Ambler in the legacy chapter, residues interact in unique ways with the natural, social, economic, political, and local environments in which glyphosates circulate. In Argentina and Sri Lanka, for example, workers and residents have apprehended glyphosate through their sometimes-unaccountable experiences of illness.

Sociologist Florencia Arancibia has studied opposition to glyphosate in Argentina intensively and has described three stages of collective protest.[103] Citizen opposition began in early 2000 with the then unknown (and now famous) Madres de Ituzaingó, who organized the first health survey, going door to door in their local village "collecting data on medical diagnoses, writing a list and drawing a map showing the location of each ill person."[104] This initiative, accompanied by regular demonstrations, attracted much media attention and generated an intervention by the Ministry of Health as well as gained help from human rights lawyers and an environmental protection foundation. These actions forced municipal-level protective regulatory changes, but the lack of city officials' enforcement of the new regulations drew new energy and new actors into the local movement.

In the second stage of mobilization, the Madres involved local physicians and made "the first attempt to develop new counter hegemonic scientific evidence through popular epidemiology, in which scientific data is produced by victims of diseases. The report, published in 2005, demonstrated about 200 cases of cancer among 5000 inhabitants."[105] This popular epidemiology initiative was replicated in other villages, towns, and cities, including Buenos Aires.

The Madres's initial success attracted attention from other activist groups, with different-yet-related aims and ways of acting, including the Grupo de Reflexión Rural, an organization that focuses "on the impact of global capitalism in the country" and opposes "the agricultural model based on the export of transgenic commodities as a new form of 'dependency.'"[106] The group launched the campaign "Stop the Spraying," which was also endorsed by other NGOs that were organizing neighborhoods locally to resist the corporate agro-economy.[107] At stake here was not only obtaining regulatory change to protect the health of workers but also a fight against "the complete agricultural model of bioeconomy."[108] The Stop the Spraying campaign broadened the political and geographical scope of public opposition to glyphosate, which now involved research collaborations between laypeople and scientists and built a strong advocacy network linking urban, suburban, and rural groups across the country.[109]

A third stage of antiglyphosate activism was initiated in 2009, when an Argentine biologist named Andres Carrasco published in a leading national newspaper a study documenting that glyphosate causes birth defects in animals. These results were not the first obtained suggesting this effect, but they were the first made available in Spanish and to a large Argentine audience, and they made a major public impact. This time, a coalition of "intellectuals and scientists, as well as international NGOs and Indigenous movements" launched a campaign not only to defend Carrasco's data and decision to publish in a newspaper rather than an academic journal but also to demand protection of the independence of science from dominant economic actors.[110] This campaign resulted in the creation of the "University Network for Public Health and Environment-Physicians of Fumigated Villages" in 2010, aimed at organizing and producing research while enhancing health care and public health.[111] These scientific and medical activists blamed the national scientific-regulatory system and sought to change it. Despite its still-marginal status within the Argentine scientific community, it has been very active in terms of both lobbying for change and producing knowledge.[112]

Although the multilevel, multiactor, and multifocused forms of resistance that have emerged in Argentina have done little to slow the expansion of GM soy agriculture,[113] the movement offers a powerful counternarrative to the "safe chemical" image promoted by industry. The movement has also made the suffering of the victims visible to Argentinian society and to the world. Yet E.U. activists against glyphosate could reproduce the images without having, in most cases, to contend with the actual bodily experiences of severe illness, disability, and death.

In Argentina, as in the E.U., opponents of glyphosate-requiring agriculture gained public support, but so far, no major regulatory change has occurred. A different constellation took shape in Sri Lanka beginning in 2014 that drew the attention of activists elsewhere. During that time, Sri Lanka became the first country to ban glyphosate after officials there decided to hold the herbicide responsible for a localized epidemic affecting the North Central Province of the country, known as Chronic Kidney Disease of Unknown Origin (CKDu). First detected in the late

1980s and early 1990s, the disease affects a predominantly rural poor population. It is difficult to determine just how many people have contracted or died from the disease over the last thirty years, though estimates suggest nearly half a million people may be affected and more than twenty thousand may have died.[114] The disease has proven perplexing to specialists. It affects mainly men over thirty, but it has also been identified more and more in women and, recently, even in teenagers and children. It prevails in a region that converted to industrialized agriculture in the 1980s to feed the country. Similar occurrences of CKDu have appeared in other industrialized agricultural regions of the world, especially Nicaragua and India. Yet the disease is not present in other Sri Lankan agricultural regions nor in many other agriculturally intensive regions in the world.[115] Its etiology remains uncertain and its status contested.[116]

In response to the disease, Sri Lankan authorities have mobilized national scientific institutions as well as World Health Organization experts to search for an environmental nephrotoxin—a toxin that specifically targets the kidney that might help explain CKDu's etiological puzzle. Several hypotheses have been suggested and explored: high levels of fluoride in the drinking water; the hardness of groundwater and the retention of heavy metals in the soil; residues of heavy metals, especially cadmium and arsenic, in and from agrochemicals; lead poisoning from aluminum eating utensils; and toxins produced by cyanobacteria. Many believe water is an important vector, and there is an increasing focus on exposure to heavy metals as a causal mechanism.[117]

In this context, a 2014 publication by a group of Sri Lankan scientists suggested that the chelating properties of glyphosate could transport heavy metals into the human body. According to the authors, "Although glyphosate alone does not cause an epidemic of chronic kidney disease, it seems to have acquired the ability to destroy the renal tissues of thousands of farmers when it forms complexes with a localized geo environmental factor ([water] hardness) and nephrotoxic metals."[118] A similar hypothesis focused on calcium contained in the Northern Central Region's hard water. According to this competing hypothesis, calcium combines with arsenic residues in fertilizers and pesticides to form calcium arsenic crystals that in turn interfere with "the antioxidant defense system in the renal tissues."[119] Both hypotheses treat CKDu as the consequence of farming practices, combined in this specific natural environment, all made mobile through the chelating properties of glyphosate pesticides. Accordingly, in addition to ongoing pesticide use, these communities are being poisoned by the accreted residues from the past made mobile and toxic in the present through a new chemical pathway.

As political scientist Asoka Bandarage and medical anthropologist Amarasiri de Silva have observed, the CKDu crisis comprises multiple issues not limited to the identification of a possible nephrotoxin.[120] Pesticides have long been of concern in Sri Lanka. The country is a heavy user of agrochemicals as they are deeply subsidized and very well marketed, and pesticides have long been associated with a vast array of damages, whether environmental or health related.[121] When the glyphosate hypothesis was formulated, concerned toxicologists, doctors, and activists, backed by the Ministry of Health, had already been campaigning to ban pesticides and to lower the use of fertilizers loaded with heavy metals. Their demands have faced strong resistance from the Ministry of Agriculture Services, agricultural extension services, large farmers (especially tea farmers whose business depends on heavy pesticide uses), and agrochemical companies.[122] Despite resistance, the Sri Lankan government has enacted policies to lower reliance on agrochemicals in their provision of technical and financial support for farmers and ban a series of pesticides, including glyphosate.[123] The 2015 ban, which attracted international attention, resulted from circumstantial and poorly implemented political bargaining during the election of a new government that year.[124] It was partly revoked in 2018, against the policy of the Ministry of Health, which promotes organic agriculture.[125]

In the meantime, the epidemic continues to strike, sustaining another type of activism. Doctors have organized patients' associations and mobilized the media to draw public attention to the inadequate care available to disease victims.[126] Compared to other countries in the region, Sri Lanka has a good health care system. Nephrology programs and clinics have been developed in the country since the 1980s with the support of the government. Yet activists still point to the lack of adequate facilities for dialysis and kidney transplants. Some patients lack access to any care for kidney disease.[127] Some activists also argue that the government and its administration have been slow to respond to the smallholder farmers, their families, and communities who have fed the country since independence and are now suffering and dying.[128]

These communities, which are out of European activists' sight, are developing their own coping strategies for the epidemic and the fear of illness and death that accompanies it.[129] Their concern is not a toxicant as such, but rather a devastating, deadly disease. The mismatch between the affliction and available information about its cause has upended local knowledge and cultural practices in these communities. Government-sponsored educational health programs have targeted the culpability of water consumption and use so that the local wells, once honored in this dry region as a sign of wealth, have come to signify disease and death.[130] New drinking, cooking, and cultural practices have emerged for avoiding well water, to the extent possible, by turning to spring water, limiting consumption, and buying bottled water when money is available. Consequently, people tend to reduce their consumption of water after long days of work, generating dehydration and in turn straining their kidneys. The relationships of vulnerable people to their environment have been altered. The widespread disease and premature deaths have led to further impoverishment as families accumulate debt to pay for care, lose income from work, and face a new form of stigma within the Sri Lankan society.

In Europe, Argentina, and Sri Lanka, the different forms of collective action that have constructed glyphosate residues as a problem have relied in one way or another on being able to set the chemical substance in a wider framework of encounters, experiences, transformations, and exchanges. Following Margaret Lock, they reveal various local biologies in which bodily and social realities "are negotiated and leveraged within historical, embodied, and political-economic relationships."[131] In other words, as Alex Nading puts it, "neither bodies nor chemicals behave the same way everywhere," in tension with the universalizing view of global health, which is also held by the chemical industry.[132] Across these locations and movements, what is apprehended, whether feared or fought or accommodated, is never solely a matter of toxicity. At stake, directly or indirectly, may be the entire socioeconomic-political system of food production, land, and water access.

Conclusion

The apprehension of residues is never a foregone conclusion, in part because residues can be notoriously difficult to find. As we have argued previously, residues have sociomaterial properties—slipperiness, unruliness, and the like—that often render them elusive. Pervasive denizens of our environment, many residues may be intuitively experienced, if not easily registered, through human senses; proof of their presence often must rely on chemical instrumentation. Our dependency on technological sensors is most obvious in human biomonitoring, in which chemical assays reveal our bodies to be ready containers for contaminant cocktails. The measurements bear witness to the accretion of residues over time and space, a personalized chemical ledger—and legacy of chemical trespass—in all of us.

Drawing on the work of Evan Hepler-Smith, biomonitoring works because chemists and toxicologists have built a standardized system that identifies and orders chemicals as discrete molecular entities—"a specific set of atoms linked by a specific network of bonds."[133] Michelle Murphy has noted, however, that such abstract formulations tell us little about how people and other life-forms actually experience residue-laden environments. Beyond the formalities of molecular bureaucracy, "the infrastructure of chemical relations that surround and make us largely resides in the realm of the imperceptible. We might feel some of our chemical relations and the pain they cause, but the fullness of our chemical relations ends up being largely conjectural."[134] Residues are there and not there, all at once, demanding and simultaneously confounding regulatory attention.

Even when negative health effects on humans or wildlife do reveal the insidious presence of chemical contaminants, that information does not always trigger environmental regulation. For instance, the authorization of glyphosate in the United States has been reviewed by the EPA but not overturned based on "post-market" observations of possible toxicity, even after the IARC designated glyphosate as "probably carcinogenic to humans" in 2015. The discrepancy is partly due to the source of data: the EPA used unpublished studies commissioned by industry, while the IARC relied on peer-reviewed publications.[135] Similarly, human biomonitoring studies undertaken by public health agencies in various nations produce a great deal of data whose interpretation and relevance to regulation remain contested. More generally, the machinery of regulatory decision-making is often intentionally distanced from the actualities of chemical exposure on the ground and does not necessarily align with public health efforts to assess exposure hazards.

To see how knowledge of chemical presence translates into regulatory *inaction*, we watched an advisory expert committee do its work. The SCOEL's setting of limit values was a social activity screened off from workplace experiences of exposure, relying on only a tiny fraction of potentially pertinent medical and scientific information. Restricting data used in formal decision-making is typical of regulatory action and is one reason that setting threshold limit values can appear arbitrary and aimed at mollifying stakeholders. David Demortain has demonstrated that the U.S. EPA's establishment of risk assessment also relied on regimented decision-making, including separate steps for identifying hazards, determining exposures, and dealing with uncertainty. While this technocratic system provided the major basis for EPA decisions in several regulatory domains for more than two decades, more recently attacks by industry and by environmental researchers have eroded its social authority.[136]

The technoscientific measurements that are central to identifying and regulating hazardous residues complicate and sometimes outright contradict efforts to legitimize human experiences of harm due to chemical violence. This conundrum gets at the heart of apprehension, which we use here to refer to both personal experience and prosthetic detection of residues. The stories of glyphosate exposures on three continents illustrate that apprehension is necessarily situated. In this case, residues are the "chelating agents" that bind to and transport entities that, in isolation, would stay put and likely go unrecognized. The unanticipated effects of glyphosate have transformed invisible exposure into visible suffering, which toxicity studies fail to predict or explain. As messengers from the past, residues spotlight prior actions and construct contexts for future decisions.[137]

Knowledge about residues is often hard-won, and not only because apprehending them may rely on costly scientific instrumentation and technical interpretation. When it comes to detecting residues, the chemical industry has clear interests in how relevant knowledge is created, shared, trusted, and interpreted. In this respect, our category of apprehension invariably relates to the large body of literature on ignorance, particularly those scholars who have documented how industry has interfered with the publication of damaging information or cast doubt on its validity if it is already public knowledge.[138] The systematic disregard for residues is part and parcel of the production and consumption of industrial chemicals, as we explore in the final chapter. Apprehension of residues, in this respect, is also political.

5

Residual Materialism

We end this book near where we began, with consideration of the Anthropocene and the problems posed by industrialization to future generations, human and nonhuman. At a time when ice sheets and islands alike are disappearing before our eyes, the ethical purpose and empirical goal of science seems no longer pointed toward advancement but, increasingly, toward adaptation, even survival. Residues offer a new way of imagining how science and technology are implicated in our rapidly changing world.

As we have described them, residues are the leftover, missing, and unaccounted-for remnants of human societies. Some residues have been here for millennia, but in the last century, they have proliferated exponentially in volume and in kind. They are material things but also social objects. In residues, we witness the combined actions and forces of economies and ecologies, regulation and innovation, waste and production. Residues aren't just what we make: they are the how, why, and where of what we make, which is why the sociomaterial properties of residues are of such critical importance to us today.

By focusing on residues and their properties of irreversibility, materiality, slipperiness, unruliness, and negative value, we have tried to reimagine our chemical environment, and how it is (mis)managed, from the ground up. Legacy, accretion, and apprehension provided distinct orientations useful for thinking through chemical domains. We used each dimension to organize empirically broad, but analytically independent, discussions of residues.

Together these three chapters represent the backbone of the book—organizationally differentiated but structurally interdependent. Legacy orients us timewise and asks us to connect residues past, present, and future. Accretion is a spatial idea that prompts questions about how residues concentrate and disperse, often in relation to human activities and the built environment as well as ecologically through landscape, topography, and hydrology. Apprehension sensitizes us to the dimension of human experience and meaning and to the sense-making work of machines and communities, whether in a lab or along a factory fenceline.

But our empirically grounded account of the pervasiveness of residues presents a puzzle. If residues are so ubiquitous, why do we not constantly notice them? What processes keep residues out of sight, both physically and socially, and how is this invisibility integral to current systems of production, consumption, and regulation? These are "big picture" questions and pursuing them, as we do through the book's final pages, leads us to flesh out a framework for considering the political implications of our project, a framework we have been calling residual materialism.

In doing so, we nod to the work of Stephen Bunker and Paul Ciccantell, whose concept of "raw materialism" focused on how physical properties of raw materials and the geography of landscapes can account for socioecological dynamics in extractive economies, transportation infrastructure, and globalization since the seventeenth century.[1] The rise of the modern chemical industry means that the physical agents now reshaping the environment are *processed*, not only raw—they are residues, in the terms we've been discussing. Hence our use of "residual materialism" to think outward from residues. In our view, a key strength of residual materialism is the flexibility it provides to scholars working at different levels of analysis, scaling up from the molecular to the institutional and back down again. The shifting perspective also suggests how studying residues can enable a broader historical analysis of the role of chemicals in industrialization, and prompt political demands for change, without requiring a totalizing view or deterministic logic. It is a way of seeing chemical residues as coproduced with the environments, institutions, and infrastructures that generate, distribute, regulate, and transform them.

We do not pretend that residual materialism provides anything like a solution to the myriad problems that residues create. Rather, we offer it as a way to track residues as sociomaterial objects even as they interact with one another, with ecological and socioindustrial systems, and with human (and nonhuman) communities. What is the intellectual payoff? To answer this question, the next sections recast the history of chemical industry and regulation with particular attention to the generation and manipulation of residues—and, not least, how they are then obscured. We end the book with some ideas about how residual materialism, focused on the conjoint sociomaterial nature of chemical environments, might better inform our responses, scholarly and activist, to the chemical Anthropocene.[2]

Scaling Up

Today, chemical production and manufacturing operate as a heterogeneous global enterprise. As evidence in preceding chapters shows, however, the technoscientific tools for conceptualizing, controlling, and managing residues flowing from the industry are part and parcel of the same approach that built and now maintains that system. This is somehow ironic, given that modern chemical industry began with a residue: coal tar, a thick and highly toxic liquid by-product of coal gasification.[3]

In the late nineteenth and early twentieth centuries, modern large-scale chemical industry owed much of its organization to local geography. In general, companies built production facilities as close as possible to the raw materials needed to produce the chemicals they marketed. The advent of new transportation infrastructures involving seaports, railroads, and later, interstate highway systems, altered the older, geographical logic of chemical production. No longer tied so tightly to local stocks of raw materials, chemical company owners instead began to see advantages in ensuring their factories' ready access to transportation infrastructure. In time, regions of dense concentration of chemical facilities developed in places like Germany's Rhine Valley and the Lyon/Rhône-Alpes region in France, where shipping, rail, and freight transport systems converged. Now raw materials and finished products alike moved in and out of production facilities more easily, efficiently, and cheaply. So did their residues.

The geographical reorganization of chemical industry ran in parallel with organizational and legal restructuring as chemical companies began morphing into the vertically integrated, supranational entities we recognize today. An important feature of the transformation was the reordering of patents following the First World War that shifted the economic power and physical concentration of the chemical industry from Germany to the United States. As a result of World War I, the Western Allies (notably the United States) confiscated German patents and initiated the rise of their own organic-chemical industry. Along with this, a change in the raw material base from coal to petroleum, enabled by new technologies for "cracking" the petroleum and obtaining a more diverse and useful set of base chemicals, occurred first and foremost in the United States.[4] The rise of New Orleans, Louisiana, and Houston, Texas, as regional hubs for chemical production along the Gulf of Mexico illustrates this point.[5] A few decades later, the Second World War created a new global order of established and emerging economies that impacted the chemical industry in two notable ways. First, the new economic order located consumption of chemical products as a defining feature of established economies, mostly located in the Global North, and relegated the messy work of chemical production to developing nations of the Global South.[6] Second, the new global division of labor untethered chemical production from the nation-state. For example, German and French chemical companies repositioned themselves beyond national borders to capture a competitive advantage within a rapidly globalizing marketplace.

These macrohistorical changes in the technological, legal, economic, and political restructuring of chemical industry had far-reaching sociomaterial impacts on the residues we are studying.

Unleashed by a complex system of extraction, transportation, and manufacturing, residues increasingly moved further and faster, and more and more they moved together, creating new ecologies of complex chemical accretions and legacies: the sociomaterial traces of late twentieth-century globalization.

The advent of early environmental and health protections only exacerbated these distinctions with laws and policies in support of some form of control of chemicals only existing in the established, increasingly consumer-based markets, while sites of growing concentrations of global production, especially in the southern hemisphere, remained largely unprotected. Global waste policies came to mirror the global flow of raw materials. Industrial centers doubly contaminate less-powerful peripheries: generating waste while extracting raw materials and then disposing of the remains after industrial production and social consumption. The advent of international markets for waste offered an opportunity to complete the global flows, ridding the wealthier consumer economies of discarded goods and sending them to regions of the global east and south whose job it became to try to assimilate these materials back into global economic cycles.[7]

Residues are not only the products of a massively scaled-up industry, but they also reflect the diversity of the chemical production chains. In the course of the last 150 years, mimics of natural products were followed by the substitution of outdated chemicals by unique synthetic molecular constructions. Innovations in chemical engineering, spurred by forces of economic expansion, competitiveness, compliance, and health, ensured a steady flow of more substances, some novel products of human chemical artifice, others human-made derivatives of the naturally occurring compounds for which they might substitute. The combination of the radical inventiveness of, for example, entirely novel substance classes (such as PCBs) with one of the basic paradigms of industrial R & D, incremental innovation, led to a whole new chemodiversity, a "zoo" of substances depending on the production chains of science and industry. Today, it is estimated that approximately seventy thousand chemicals are actively manufactured worldwide.[8]

Although this is a staggeringly high number, it is perhaps more interesting to note that the chemicals manufactured with the highest production volume have remained largely unchanged for many decades, if not longer: sulfuric acid, ammonia, chlorine, ethylene, and benzene among them.[9]

The tremendous growth and diversification of chemicals' production and use since World War II have created a paradox of enormous economic fortune combined with an uncertainty and trepidation surrounding the industry itself. And as *chemical* has come to refer to an ever-expanding assortment of molecules and products for those that make and deal with them, the word has become more singular in the vernacular of popular culture, often associated with (or assumed to be associated with) expressions of danger, fear, risk, and unnaturalness. This disconnect is more than semantic; it perpetuates the systems of production and control that operate largely out of sync with one another, a disconnect we have signaled with the apprehension of residues. While an overarching regulatory category has been generated by the use of the notion of "chemicals," the regulatory tools, instruments, and institutions designed to oversee these materials have "cracked" into many parts, like the substances themselves and like the categories of substances. Put a bit differently, residues regularly escape the categories regulators and scientists have built to tame and contain them.

As regulations are fragmented, overly specific, or narrowly focused, they lose track of effects that are, by definition, long-term, wide-ranging, and diverse. One of the most urgent tasks identified by the current community of ecotoxicologists is the systemic nature of chemical pollutants. Two recent scientific surveys, focusing on Europe and Latin America, respectively, point to the urgent gaps in our knowledge of chemical contamination.[10] The spottiness of the knowledge produced by chemical effects research is a direct result of a model of economic growth that privileges production over health and immediate utility over considerations of material life cycles. Chemical residues are the lingering underside of this economic model, but they are also the markers of residual modes of production, characterizing an industry that celebrates and valorizes innovation while continuing to mass-produce commercial chemicals, intermediates, and wastes whose toxicities have been known for decades. The materiality of residues helps us comprehend the industrial system by tracking its remainders. Different chemicals tell different stories—PCBs in waterways, asbestos on the clothes of workers, glyphosate in the bodies of farmers—yet they also point to the more systemic features that characterize industrialization. In all this, our current systems perpetuate the creation of unaccounted-for residues—through purposeful design, fractured oversight, negligence, and individual or collective wrongdoing.[11]

Agency and Invisibility

Our framework of residual materialism relies on the role that scientists and engineers exert on their environment through the materials they make. A hallmark of laboratory chemists over the course of the modern period, in both industry and academia, has been the synthesis of new substances. In this endeavor, most substances are not products themselves but intermediates—steppingstones—in reaction chains leading to other products and consumer goods. Making use of new and old substances in ever-novel combinations or sequences has provided both the basis for the chemical enterprise and the increasing diversity of the human-built world. However, chemical reactivity and malleability do not stop at the boundaries of creativity and intention. Chemicals continue to act, to exert their agency, in unforeseen ways, creating new realities after they have left the chemists' laboratories and manufacturing plants, the homes of consumers, or the detritus of a razed building. Familiar examples include DDT, which not only helped to combat malaria but also killed birds and other animals, and CFCs (chlorofluorocarbons) that were applied in spray cans and as refrigerants but in the upper atmosphere caused the destruction of ozone.

Such agency, often not immediately recognizable and frequently brought about by material metamorphoses, is the single most important defining feature of residues. Chemical residues remind us how materials exist both as products and as agents of change—ontological, epistemic, and social. Whether emerging from academic or industrial laboratories, molecules are designed to perform in known and predictable ways as both tools and agents. Once mass-produced in chemical plants and released into the world, however, the agency of chemical residues can belie common assumptions of technological control.

Yet the issue is not only one of the subversive agency of residues. Running concurrent with industrial growth and scientific innovation is a set of kindred processes operating to move and bury waste, to reduce and eliminate value, and to hide leftovers in plain sight. Though clearly part of industry, these processes of "invisibilization" define residual materials away through economic theory, legal practice, scientific study, and regulatory oversight.[12] Importantly, the relegation of residues to the sidelines is not accidental. To pay attention to waste, contamination, and environmental degradation would interfere with prioritizing innovation, production, consumption, and economic growth. Moreover, due to the centrality of chemicals in economic life, all industrial sectors from agriculture to armaments are complicit.

Hiding chemical and other residues is an active process that can take as much effort as extracting the raw materials in the first place.[13] This process may be intentional, but activities that marginalize residues also result from compliance with bureaucracy, municipal waste removal, and simple inattentiveness. We remind readers that consumption is not an end stage, as many consumers are barely aware of the residues they generate. Industries and institutions ensure unawareness. Economics externalizes these costs, law socializes their risks, and science identifies limits on what is and can be known. For example, laws regulating chemical exposure by the concept of threshold limit values permit regulators to ignore residues falling below a certain specified level. In other respects as well, our regulatory systems mostly hold residues in a state of "in-between," oscillating between existence and disappearance. In this way, residues remain physically present even when systems of oversight render them unrecognized or unrecognizable. With residual materialism, we seek to bring them back into view and make them objects of concern.[14]

Apprehension of residues is a fraught enterprise, with enormous economic and political stakes; many people, organizations, and governments would prefer that societies' collective ignorance of residues continue largely unchallenged. Consider similar societal processes of invisibilization: So-called cloud computing hides the actual labor and massive environmental costs entailed by moving data storage to remote sites, accessible by wireless technologies. And it is not only objects that are hidden or devalued in this way. As demonstrated by Black Lives Matter in the United States, social movements are critical in challenging the invisibilization of Black Americans and of the routine violence against them. Visibility is not (only) determined by physical presence but also by social recognition.

Residual materialism captures how residues persist materially and dynamically in the environment even as they socially disappear. As we have shown in chapter 2, when communities are expected to live with industrial waste or long-manufactured substances are not tested for toxicity, residues in the environment are socially effaced, and the chemical legacies of the past are disconnected from the future. Similarly, in chapter 3, we have described how the accretion of residues often involves their failure to stay within the systems that we have built to channel and contain them. Where residues go and whether and how they accumulate are structured by their interactions with ecological and social systems.

Thus legacy and accretion ensure that residues continue to play a role in our midst. This emphasis contrasts with the epistemological approach first favored by the environmental sciences or by STS scholars who study environmental ignorance.[15] By framing ignorance predominantly as an epistemological problem—a problem of absent knowledge—students of agnotology and ignorance studies more generally sometimes miss the sociomaterial and ontological dimensions of what is not known. For instance, in post-Katrina New Orleans, ignorance took sociomaterial form as city blocks left unsampled by the U.S. Environmental Protection Agency.[16] The disconnect between the epistemology of residues and their ontology is what makes residues so effective and at times dangerous. The uncounted victims of glyphosate poisoning, the mass die-offs of insects on the German plains, and the legacy contaminants lurking beneath city streets—by using residual materialism, we seek to resurrect the object-nature of the unknown.

Facing the Chemical Anthropocene

Residues are the chemical currency of the Anthropocene.[17] The presence of chemical residues, such as radioactive isotopes, PCBs, or plastics, serves as an argument for the deep and lasting impact of industrial activities on the earth. They also serve as an important reminder that, while nearly all of today's debate on anthropogenic environmental harm centers on climate change, we should not reduce the crisis to one of greenhouse gas emissions. Other, damaging effects deserve critical attention as well. Fossil "fuels" power the modern world, but they also serve as feedstock for most of the synthetic materials in use today: fertilizers and pesticides in agriculture; plastics, fibers, dyestuffs, and other materials for household items and clothing; chemicals used as pharmaceuticals and in electronic devices—to name just the most important applications. In addition, inorganic chemicals, metals, and mineral-based materials have experienced an enormous increase in production and use as well. Alongside the carbon dioxide and other emissions generated by burning fossil fuels, all these manufacturing activities leave their own material, sometimes toxic, remnants.

Only when we take the full scale and scope of residues into account—not just the greenhouse gases involved in climate change and the CFCs damaging the ozone layer—can we recognize the complex spectrum of anthropogenic impacts on our globe. The planetary spread of residues appeared on the scene well before the advent of capitalism and did not primarily involve fossil fuel-based CO_2 emissions. Roman smelting of ores caused an increase in atmospheric lead and, many centuries later, Spanish mining activities for gold and silver in sixteenth-century South America released atmospheric mercury. Capitalism and the colonial projects required by its logic of accumulation did kick the generation of residues into higher gear.[18] The First Industrial Revolution between 1760 and 1830 brought the widespread use of coal for energy and iron and steel production. The Second Industrial Revolution made coal and later oil the feedstock material for an ever-widening menu of economically useful, carbon-based materials. Most recently, the so-called Great Acceleration impelled by Cold War dynamics and technologically enabled access to "cheap oil" also unleashed a dramatic increase in the scale and scope of fossil resource use aimed at developing molecularly unique synthetics.

Viewing the Anthropocene through the lens of residual materialism offers new ways to think through chemical environments. For one, it makes clear that we do not get very far by treating chemical residues as little more than passive confirmation of a new geological age. As we have argued throughout this book, residues carry properties that are simultaneously material and social. They are not merely inert material evidence, the by-products of expansive capitalist machinery, but active sociomaterial agents making and remaking the world in expected and unexpected ways. Our goal is to understand residues' world-making powers.

Indeed, in shifting attention from the scientific practice of researchers to the socioecological propensities of residues, we should remain mindful of the absences, inactions, and nondecisions that leave residues out, uncounted, and unaccounted for. An obvious example is the industrial production model that treats residues as externalities; another is the regulatory principle, dominant in the U.S. context, which finds most commercial chemicals innocent until proven guilty.[19] In both realms, residues are afterthoughts, often unnoticed and inhabiting socioecological spaces beyond the reach or attention of actor-networks.[20] Residues help us spotlight spheres of intentional neglect and reckon with the magnitude of undone science on the environmental impacts of chemicals. The blinders put on regulatory knowledge-making mean that industrial production and pollution often continue largely unabated. Or, when regulatory systems seem to work, more likely they are actually shifting pollution from one part of the biosphere to another.

A case in point is in the United States, where federal laws circa 1970 regulating air and water quality had the perverse impact of increasing industrial waste accumulation on land—the stuff had to go somewhere, so companies diverted pollution streams from rivers and smokestacks on top of or below ground, into dumps, pits, lagoons, ponds, and wells.[21]

In helping us see the unseen, residual materialism also serves as an antidote to brute technocratic approaches to fixing the planet, notably arguments for geoengineering systems that are gaining momentum in earth system sciences and policy circles.[22] Following residues reveals why engineering our way out of the climate crisis will not do. Technological solutions to environmental problems inevitably generate their own residues. The hydrometallurgical processes that drive China's solar panel industry also created Baotou's massive toxic lake.

And yet, residual materialism's main analytic value, we think, derives from the fact that residues have never attracted the kinds of scientific and engineering capital that magic bullet projects for geoengineering or alternative energy infrastructure already do. Residues are and always have been a marginal topic—for scientists and engineers as well as for environmental STS scholars and kindred folk. Yet we are convinced that following residues can lead us to the heart of the problem. It can show us how capitalism works from the inside out, in the cracks and leaks inherent to macroeconomic systems of production, regulation, and exchange. As Michelle Murphy has argued, "governing industry for the sake of the macroeconomy is a set up that produces molecular material 'waste' emissions as outside of the calculation of value." She describes, in broad strokes, the accumulation of residues' negative value that too often concentrates in places of human neglect, where "decades of purposeful polluting" create "diminishing conditions for the unvalued many."[23]

Residual materialism centers the out-of-the-way spaces—and people—that capitalism neglects.[24] It is one reason critical analysis of capitalism or the "Capitalocene" alone will not suffice.[25] Abstractions will not help us address—or even apprehend—the pervasive realities of chemical contaminations; our focus on residues has convinced us that specificity and locality matter, in our activism and in our scholarship. We must also reckon with the hard truth that there is no going back to a world without anthropogenic alteration of the planet; a world without complex industrial systems of material production, consumption, and disposal; a world without a residual past, present, and future.[26]

So in the spirit of "staying with the trouble," as Donna Haraway has phrased it, we have tried to work from the concrete, but often neglected, remnants of our chemical world, locating ourselves clearly within it, as participants, not neutral observers.[27] In part, this is a matter of moral concern: the seven of us are not only witnesses to these processes but also contributors to the everyday production of our shared residual reality. We inhabit a permanently altered biosphere as scholars but also as living beings and members of diverse national, civic, and professional communities.

From the beginning, we have embraced this project as an opportunity for engaging others who are frustrated, intellectually and politically, with the current frameworks for understanding chemicals in the environment. In addition, *Residues* is a vehicle for thinking out loud with our colleagues in the history of science and technology and STS, and in particular those working in environmental STS, environmental history, and environmental sociology. This community of scholars bears a special and growing burden.[28] As practitioners of a niche academic craft and occupants of a planet in general crisis, our work must internalize and build from an apparent contradiction: the materiality of socially constructed nature. Taking the conjoint social and material nature of the problem seriously means that we must work with science even as we marshal critiques and even as we challenge institutions that produce, circulate, and sometimes suppress relevant knowledge.

Residues connect. These shape-shifting world travelers depend on the link between the social and the material for their very existence. To track them, we need combined approaches of the natural sciences, the social sciences, and the humanities to research residual states of matter. Turning our focus toward residual materialism does not automatically enable solutions to our myriad environmental problems. But it can help us see more clearly those parts of the world that our current systems are designed to overlook and choose instead to find and face them.

Notes

Chapter 1: Residue Properties

1 Ilgen, "Better Living through Chemistry," 650. Ilgen's phrase is summarizing J. D. Bernal's observation: "The chemical industry . . . has become the central industry of modern civilization, tending, because of its control over materials, to spread into, and ultimate incorporate, older industries such as mining, smelting, oil-refining, textiles, rubber, building, and even, through its concern with fertilizers and food processing, agriculture itself." Bernal, *Science in History*, 823.

2 Buccini, *Global Pursuit*, xiii.

3 Buccini, xiv–xv.

4 Geyer, Jambeck, and Law, "Production, Use, and Fate."

5 Bowker and Star, *Sorting Things Out*.

6 Egan, "Chronicling Quicksilver's Anthropogenic Cycle"; Selin and Selin, *Mercury Stories*.

7 Nisbet and LaGoy, "Toxic Equivalency Factors."

8 Kellow, *International Toxic Risk Management*.

9 These information gaps are problems for industry as well. Scruggs and Ortolano, "Creating Safer Consumer Products."

10 "Available" here means in the public record. Environmental Defense Fund, *Toxic Ignorance*.

11 Denison, *Orphan Chemicals*.

12 Roberts, "Unruly Technologies."

13 Douglas, *Purity and Danger*.

14 This literature is growing rapidly, but for a sense of its breadth, see, for example, Boudia, "Managing Scientific and Political Uncertainty"; Brown, *Toxic Exposures*; Cranor, *Regulating Toxic Substances*; Jasanoff, *Fifth Branch*; Krimsky, *Hormonal Chaos*; Langston, *Toxic Bodies*; Markowitz and Rosner, *Deceit and Denial*; Pellow, *Resisting Global Toxics*; Proctor, *Cancer Wars*; Sellers, *Hazards of the Job*; and Shostak, *Exposed Science*.

15 Brickman, Jasanoff, and Ilgen, *Controlling Chemicals*. Among those STS scholars who have focused on the chemical industry, we especially note Fortun, *Advocacy after Bhopal*; Casper, *Synthetic Planet*; Allen, *Uneasy Alchemy*; Murphy, *Sick Building Syndrome*; Henry, *Amiante*; Ross and Amter, *Polluters*; Vogel, *Is It Safe?*; Boullier, *Toxiques légaux*; and Jarrige and Le Roux, *Contamination of the Earth*. There is also substantial scholarship on chemicals regulation in various national contexts, which we will not attempt to cite here, including work by sociologists, political scientists, legal scholars, and environmental and labor historians. For

examples of work on transnational regulation of chemicals see endnote 24 in this chapter.

16 On materiality in STS and social theory, see Bennett, *Vibrant Matter*; and Law, "Materials of STS." On infrastructure in STS, see Bowker and Star, *Sorting Things Out*; and Slota and Bowker, "How Infrastructures Matter."

17 Gabrielle Hecht's theorization of residue as a problem of the governance of contaminants in mining is consonant with, but narrower than, our own. See Hecht, "Residue." For an earlier version of our approach to the concept, see Boudia et al., "Residues."

18 Barad, *Meeting the Universe Halfway*.

19 Latour, *Reassembling the Social*, 12; Casper, *Synthetic Planet*.

20 Lexico, s.v. "residue," 2021, https://www.lexico.com/en/definition/residue.

21 See definition 4 of residue in the Oxford English Dictionary online: https://www.oed.com.

22 Jas, "Adapting to 'Reality.'"

23 Casper, *Synthetic Planet*; Creager, *Life Atomic*.

24 The "transboundary" nature of chemical pollution made it an issue of international negotiation and regulation by the 1970s. Selin, "Drawing Lessons"; Long, *International Environmental Issues*; Pallemaerts, *Toxics and Transnational Law*; Selin, *Global Governance*; Eriksson, Gilek, and Rudén, *Regulating Chemical Risks*; Rothschild, *Poisonous Skies*.

25 Cowles, *Scientific Method*, 18. For additional development of this idea, see Cowles, "Review of *Life Atomic*."

26 The classic statement is Bunker, "Modes of Extraction." See also Bunker, "Staples, Links, and Poles."

27 Spears, *Baptized in PCBs*; U.S. Congress, Senate, *Implementation*.

28 Serres, *Malfeasance*, 42.

29 Reinhardt, "Regulierungswissen und Regulierungskonzepte." On regulatory inaction, see Bachrach and Baratz, "Two Faces of Power"; and Henry, *Ignorance scientifique*.

30 Shotwell, *Against Purity*; Tsing, *Mushroom*; Gramaglia, "Saltkrake."

31 Beckert, *Imagined Futures*.

32 McNeill and Vrtis, *Mining North America*.

33 Khetan, *Endocrine Disruptors*.

34 Chen and McKone, "Chronic Health Risks."

35 Tainter, "Global Change."

36 Bohme, *Toxic Injustice*; Daniel, *Toxic Drift*; Harrison, *Pesticide Drift*.

37 Beamish, *Silent Spill*.

38 Dale et al., "Modeling Nanomaterial."

39 Walker, *Killing Them Softly*.

40 EFSA Panel on Contaminants in the Food Chain, "Presence of Microplastics and Nanoplastics."

41 Richter, Cordner, and Brown, "Non-stick Science."

42 These are called secondary pollutants.

43 Seiler and Berendonk, "Heavy Metal."

44 Crooks, "BP Draws Line." This figure does not include costs of civil penalties, environmental damages, criminal fines, and compensation to businesses and individuals.

45 Calculated for inflation at August 2016 dollars. McMahon, "Historical Crude Oil Prices."

46 Mosbergen, "New Plastic Garbage Patch." Scientists have identified five such massive patches in the world's oceans.

47 United Nations Environment Programme, *Global Report.*
48 Cf. Cordner, Richter, and Brown, "Chemical Class Approaches."
49 Baccini and Brunner, *Metabolism;* Dijst et al., "Exploring Urban Metabolism"; Moore, *Anthropocene or Capitalocene?;* Frickel et al., "Undone Science"; Hess, "Niche-Regime Conflicts."
50 We do not in the least deny the value of this materialist approach, which traces back to Karl Marx; see Foster, "Marx's Theory of Metabolic Rift."
51 Jasanoff, *States of Knowledge.*
52 Haraway, "Cyborg Manifesto"; Martin, "Anthropology and the Cultural Study of Science."
53 Gross and McGoey, *Routledge International Handbook.* On this large literature, see the "Science and Ignorance" database project: https://caphes.ens.fr/science-and -ignorance-online-bibliographic-database/.
54 Frickel et al., "Undone Science."
55 Colborn, Dumanoski, and Myers, *Our Stolen Future;* Buhs, *Fire Ant Wars;* Zierler, *Invention of Ecocide;* Suryanarayanan and Kleinman, "Be(e)coming Experts"; Puig de la Bellacasa, *Matters of Care;* Foucart, *Et le monde devint silencieux.*
56 In this vein, environmental and STS scholars are not only working with communities and science allies impacted by industrial activities (Jalbert et al., *ExtrACTION;* Hoover, *River;* Liboiron, Tironi, and Calvillo, "Toxic Politics") but also developing new ways of studying the socioeconomic impacts of industrial pollution. Liboiron, "Redefining Pollution"; Senier et al., "The Socio-exposome"; Wylie, Shapiro, and Liboiron, "Making and Doing"; Wylie, *Fractivism.*
57 For some orientation to this literature, see the contributions to Moore, "Anthropocene"; Trischler, "Anthropocene." On the term, see Crutzen and Stoermer, "'Anthropocene'"; Crutzen, "Geology of Mankind." As proposed by Crutzen and Stoermer in 2000, the Anthropocene refers to a period of human-induced changes to Earth's biogeochemical systems. It is meant to replace the Holocene as the current geological era, and Crutzen and Stoermer suggested that the late eighteenth century provided a useful starting date consonant with James Watt's improvement of the steam engine in 1784. Among geologists, debate continues about when the Anthropocene begins. Some argue for ten thousand years ago with the rise of large-scale agriculture; others argue for 1950, with the commencement of nuclear testing.
58 Frickel, *Chemical Consequences;* Landecker, "Antibiotic Resistance."
59 Pierson, *Politics in Time.*
60 Douglas, *Purity and Danger.*
61 Here we draw on the work of Shapiro, Zakariya, and Roberts, "Wary Alliance," and take inspiration from Murphy's "regimes of imperceptibility" and ways of apprehending chemical exposures in *Sick Building Syndrome,* especially chapters 4 and 5.
62 Hobsbawm, "Social Function"; Hobsbawm, "Introduction."
63 Shotwell, *Against Purity;* Tsing, *Mushroom.*

Chapter 2: Legacy

1 U.S. Environmental Protection Agency, "ACE." While this study is based in the United States, contamination from PCBs and other persistent organic pollutants is global.
2 Spears, *Baptized in PCBs,* 1–4; U.S. Congress, Senate, *Implementation,* 3–12.
3 Monsanto Chemical Company took over the Swann plant in Alabama, becoming the sole U.S. manufacturer by 1935, and the main producer worldwide for the

ensuing decades. For the EPA press release on its PCB manufacturing ban, see U.S. Environmental Protection Agency, "EPA Bans PCB Manufacture."

4   Sinkkonen and Paasivirta, "Degradation." The various specific chemical forms of PCBs are referred to as congeners, and the degradation varies significantly depending on how highly chlorinated the compounds are.

5   U.S. Congress, Senate, *Implementation*, 3–4.

6   Aidala, "Toxic Substances Control Act," 11.

7   Nora, *Les lieux de mémoire*. Published in English as Nora, *Realms of Memory*.

8   See, for example, Koselleck, *Vergangene Zukunft*. Published in English as Koselleck, *Futures Past*. Wajcman, *Pressed for Time*; Zerubavel, *Time Maps*.

9   This quotation and those following are from residents of Ambler, Pennsylvania, interviewed in 2012–2013, as part of Resources for Education and Action for Community Health in Ambler (REACH Ambler) through a partnership between University of Pennsylvania's Perelman School of Medicine and the Science History Institute, on whose website the interviews are archived. This project was funded through the National Institutes of Health R25-OD010521-01 from July 2012 to June 2018. At the time of the study, Beth Pilling was an administrator for the Montgomery County (Pa.) Open Space Program and attended many of the community meetings described below. Pilling, REACH Ambler, 8.

10 Center of Excellence in Environmental Toxicology, "Mapping Ambler."

11 The asbestos waste in Ambler came to be defined by two Superfund sites separated in space by Butler Pike (a main artery through town) and in time by nearly two decades. The "Piles" included the area in and around the "White Mountains" and contained areas of housing and public space in close proximity to the main factory and the "boiler house" where the asbestos and concrete slurries were produced. The "Bo-Rit" site located just north of Butler Pike includes a reservoir and an open, sprawling parcel of land that had been used as a recreation site by the adjacent, and predominantly African American, population. The closing of these sites and the establishment of Superfund designations restricted access to these few outdoor recreational spaces that had been a part of the community for generations. Center of Excellence in Environmental Toxicology, "From Factory to Future."

12 Adams, REACH Ambler, 10.

13 The National Priorities List is compiled and managed by the EPA in order to prioritize and direct Superfund resources for cleanup of particular sites.

14 Conner, REACH Ambler, 21.

15 Pilling, REACH Ambler, 22.

16 Boccuti, REACH Ambler, 13.

17 Cooke-Vargas, REACH Ambler, 21.

18 Cooke-Vargas, 22.

19 Weeks, REACH Ambler, 28.

20 Zaharchuk, REACH Ambler, 11.

21 Zaharchuk, 17–18.

22 Summit Realty Advisors, LLC, "Ambler Boiler House."

23 McDonough, REACH Ambler, 20.

24 McDonough, 20.

25 McDonough, 20.

26 Comparison cases of communities that struggle to define their future in the face of industrial-scale legacy residues include Seveso (Italy), Flammable (Argentina), and New Sarpy (United States). Centemeri, "Investigating"; Auyero and Swistun, *Flammable*; Ottinger, *Refining Expertise*.

27 For scholarship examining this claim in more detail, see, for example, Pestre, *Le gouvernement*; and Pallemaerts, *Toxics and Transnational Law*.

28 We will use the term *system* to refer to the complex of regulations, even as we acknowledge that it is not a coherent or rational entity, as the term *system* may suggest. Its legal and administrative parts are, however, interconnected even as they may be in tension with or even contradict each other. Roberts, "Unruly Technologies"; Jas, "Gouverner les substances."

29 Davies and Davies, *Politics of Pollution*, 10–11.

30 Davies and Davies, 26–60.

31 The U.S. government participated in the concern about environmental cancer. In 1968, the National Cancer Institute launched a "Plan for Chemical Carcinogenesis and the Prevention of Cancers." During the subsequent years, this agency publicized an estimate that as much as 90 percent of human cancer was due to environmental agents. Library of Congress, *Legislative History*, 532, 539. This figure is usually attributed to Higginson, "Present Trends." Boyland went one step further and stated that the 90 percent was due to chemical components. Boyland, "Correlation."

32 The major laws include the following: National Environmental Policy Act, 1970; Clean Air Act, 1963; Clean Water Act, 1972; Federal Insecticide, Fungicide, and Rodenticide Act, major rewrite in 1972; Safe Drinking Water Act, 1974; Resource Conservation and Recovery Act, 1976; and Toxic Substances Control Act, 1976. Davies and Mazurek, *Pollution Control*, 12.

33 Davies and Mazurek, 16.

34 Corn, *Protecting the Health*; Noble, *Liberalism at Work*.

35 "Toxic Substances Control Act of 1976," Code of Federal Regulations, 15 U.S.C., §§2601–2629 (1976).

36 J. Clarence Davies was working in the Council for Environmental Quality, which was established within the Executive Office of the U.S. President as part of the National Environmental Policy Act in 1969. Davies, "Toxic Substances Control Act."

37 Ingle, "Background," 2.

38 O'Reilly, "Torture by TSCA," 43.

39 O'Reilly, 43; Creager, "To Test."

40 In fact, the United States continued to produce a smaller amount of PCBs for export, though production was also happening in Europe and is now primarily taking place in Asia.

41 Following World War II, the Department of Defense began storing PCB waste on Guam, a U.S. territory in the South Pacific. James Aidala, who worked at the EPA during the late 1970s, recalled being called nearly every day by the political representative from Guam about getting PCBs off the island. And every day, Aidala would have to tell him that the EPA was still working on regulations and procedures for dealing with PCBs. Regulation often produces inaction, at least initially—but possibly indefinitely. Aidala, "Toxic Substances Control Act," 12. On the tendency for regulation to stymy action, see Henry, *Ignorance scientifique*.

42 O'Reilly, "Torture by TSCA," 43.

43 Gaynor, "Toxic Substances Control Act," 1151.

44 O'Reilly, "Torture by TSCA," 43.

45 On the lack of innovation in the pharmaceutical industry, see Sismondo, *Ghost-Managed Medicine*.

46 The law actually authorized EPA to specify toxicity test guidelines, but in the late 1970s and the early 1980s, the MCA (in 1978 renamed CMA) successfully

countered implementation of agency testing specifications, arguing that this constituted statutory overreach. Davis Le Blanc, "Initiatives."

47 Office of Technology Assessment, *Information Content*, 6.

48 Silbergeld, Mandrioli, and Cranor, "Regulating Chemicals." In the absence of hard toxicology data, the EPA turned to Structure-Activity Relationship analysis, looking for chemicals of concern via structural similarities to known toxic substances. Boullier, Demortain, and Zeeman, "Inventing Prediction."

49 Schierow, *Toxic Substances Control Act*.

50 Brickman, Jasanoff, and Ilgen, *Controlling Chemicals*, 142; Stadler, "Corrosion Proof Fittings."

51 On risk assessment, see Boudia, "Managing Scientific and Political Uncertainty"; and Demortain, *Science of Bureaucracy*.

52 Geiser, *Chemicals without Harm*, 1.

53 Biles, "Harmonizing."

54 The OECD came into being in 1961 (from a treaty signed in 1960) as an outgrowth of the Organisation for Economic European Co-operation, which had been established in 1948 to administer American aid to European countries as part of the Marshall Plan. The OECD was aimed at integrating European national economies, particularly through addressing nontariff barriers such as product regulations. Non-European countries began joining the OECD, beginning with Japan in 1964.

55 Long, *International Environmental Issues*.

56 Rothschild, "Burning Rain"; Rothschild, *Poisonous Skies*.

57 Lönngren, *International Approaches*, 194–195.

58 Lönngren, 199.

59 Lönngren, 203.

60 On the role of business interests in shaping OECD decisions, see Lanier-Christensen, "Creating Regulatory Harmony."

61 For an analysis of the process that led to REACH Regulation, see Selin, "Coalition Politics." For information on the regulation of chemical substances at European level before the REACH regulation, see Heyvaert, "Reconceptualizing Risk Assessment."

62 Several categories of chemicals are specifically exempted, such as those used in medicinal products and foodstuffs, polymers, naturally occurring substances, and radioactive materials. In addition, only substances imported into the European Union in quantities of more than one ton per annum must be registered. Williams, Panko, and Paustenbach, "European Union's REACH Regulation"; Boullier, *Toxiques légaux*.

63 Boullier, "Évaluer."

64 On the expectations associated with REACH, see Hansen and Blainey, "REACH"; and Heyvaert, "No Data, No Market."

65 Oertel et al., *REACH Compliance*, 23.

66 Boullier, *Toxiques légaux*.

67 Boullier.

68 Green Science Policy Institute, "Flame Retardants." See also Cordner, *Toxic Safety*.

69 Green Science Policy Institute, "September 2019."

70 Murphy, "Alterlife"; Hepler-Smith, "Molecular Bureaucracy."

71 Davison, "Ethiopia's Deadly Rubbish."

72 *Economist*, "Mountain of Rubbish."

73 Cohen, *Consumer's Republic*; Strasser, *Waste and Want*.

74   Sharma, "Delhi's Air."
75   For example, airborne dioxin and ultrafine particles generated by waste inciner-
     ation continue to be a matter of concern for scientists and for the communities
     surrounding these incinerators. National Research Council, *Waste Incineration*;
     Moll-François, "Problématiser les contaminations."
76   Karuga, "Largest Landfills."
77   Geyer, Jambeck, and Law, "Production, Use, and Fate."
78   Jambeck et al., "Plastic Waste Inputs."
79   Liboiron, "Redefining Pollution." See also Gabrys, Hawkins, and Michael, *Accu-
     mulation*; and Monsaingeon, *Homo detritus*.
80   Moore, "Trashed."
81   The classic text in the United States is Bullard, *Dumping in Dixie*. See also Pellow,
     Garbage Wars; and Hooks and Smith, "Treadmill."
82   Pellow, *Resisting Global Toxics*; Little, *Burning Matters*.
83   Barles, *L'invention*; Melosi, *Sanitary City*; Luckin, "Pollution"; Cooper, "Waste."
84   Barles, *L'invention*.
85   Tarr, *Search*.
86   Newman, *Love Canal*; Organisation for Economic Co-operation and Develop-
     ment, *Trade Measures* (on the "Seveso affair," see 98 and 172).
87   Melosi, *Garbage in the Cities*.
88   Tauveron, *Les années poubelle*.
89   Melosi, *Garbage in the Cities*.
90   Organisation for Economic Co-operation and Development, *Extended Producer
     Responsibility*.
91   Crooks, *Giants of Garbage*.
92   Pellow, *Resisting Global Toxics*; Clapp, "Toxic Exports."
93   Interpol, "Hazardous Materials Seized."
94   Sze, *Noxious New York*. On the rehabilitation project, see New York City Depart-
     ment of Parks and Recreation, "Freshkills Park."
95   Maloney, "Legislative History." Beyond the court cases over liability, a tax on
     petroleum and chemical industries to create the "Superfund" for cleanup was
     suspended in 1995 due to its unpopularity with industry.
96   Fagin, *Toms River*.
97   Burton, Kates, and White, *Environment*, 23; Knowles, "Learning from Disaster?," 775.
98   Frickel and Vincent, "Hurricane Katrina."
99   Pritchard, "Envirotechnical Disaster."

Chapter 3: Accretion

1   Clark, *Radium Girls*; Fellinger, "Du soupçon"; Harvie, *Deadly Sunshine*.
2   Zalasiewicz et al., "New World."
3   Steingraber, *Living Downstream*.
4   Hecht, *Being Nuclear*; Brown, *Manual for Survival*.
5   GZA Environmental, "Site Investigation Report."
6   GZA Environmental.
7   Specifically, the tests revealed concentrations of benzene, ethylene, toluene, xylene,
    and the gasoline additive methyl tert-butyl ether, well above regulatory limits.
8   These and other activities associated with the regulatory response to the
    gasoline release are summarized in GZA Environmental, "Site Investigation
    Report—Addendum."

9 Phase I preliminary assessments gather historical and other information about past land uses and physical site conditions. Phase II site assessments involve testing (air, soil, groundwater) to characterize the presence of environmental hazards and level of contamination. See "Comprehensive Environmental Response, Compensation, and Liability Act of 1980," *Code of Federal Regulations*, 42 U.S.C. §§ 9601–9657 (2018).

10 U.S. Environmental Protection Agency, "Leaking Underground Storage."

11 In the U.S. context, these other actors typically include property owners or other "responsible parties" and an array of private analytical laboratories, environmental consultants, and engineering firms.

12 U.S. Environmental Protection Agency, "Citizen's Guide."

13 Frickel, "Cities Are Subterranean Disasters."

14 "Comprehensive Environmental Response, Compensation, and Liability Act of 1980," *Code of Federal Regulations*, 42 U.S.C. §§ 9601–9657 (2018).

15 It follows from this line of argument that less knowledge is produced in low-income, blighted areas where properties values are lowest but where risks—and thus need for site assessments—may be higher.

16 U.S. Bureau of Labor Statistics, "Quarterly Census of Employment and Wages"; National Association of Convenience Stores, "Key Facts about Fueling." NACS statistics are updated quarterly.

17 Bell et al., "Automated Data Extraction."

18 Bell et al. Note: the statistics are based on total number of sites and so underestimate the number of redeveloped sites. There are almost certainly more than eighty thousand former gas station sites hidden within the dataset.

19 Frickel and Elliott, *Sites Unseen*. As described in this study, "socio-environmental succession" involves two additional processes: Residential churning, or the continual movement of social groups into, around, and out of different urban neighborhoods, as when an ethnic Polish neighborhood gives way over time to an influx of Mexican and other Latin American immigrants. And risk containment, when regulatory practices adopt a "worst first" strategy of site investigation, testing, and remediation. The three processes are interrelated, each one reinforcing the other two.

20 Frickel and Elliott.

21 Data collection for this project involved exhumation of historical data from annual manufacturing directories for Rhode Island, 1953–2016. The computational techniques used to extract the textual data are described in Berenbaum et al., "Mining Spatio-temporal Data." The first analysis of the environmental justice consequences of industrial churning in Rhode Island's urban core are described in Marlow, Frickel, and Elliott, "Do Legacy Industrial Sites."

22 This is a key finding from our earlier studies: deindustrialization has not meant appreciably fewer manufacturing sites. While manufacturing jobs have disappeared, manufacturing sites continue to churn at nearly the same rate. Frickel and Elliott, *Sites Unseen*.

23 Erikson, *New Species*.

24 For a few historical overviews of asbestos in industry and public health, see Castleman and Berger, *Asbestos*; McCulloch, *Asbestos Blues*; Tweedale, *Magic Mineral*; Maines, *Asbestos and Fire*; and Bartrip, *Beyond the Factory Gates*.

25 Frank and Joshi, "Global Spread."

26 According to the World Health Organization, 125 million people worldwide are exposed to asbestos in the workplace, and every year around 107,000 people die

of asbestos-related diseases. Stayner, Welch, and Lemen, "Worldwide Pandemic."
Financially, Jasanoff and Perese report that the total cost of asbestos-related
disease compensation in the United States alone is estimated to be $200–$275
trillion. Jasanoff and Perese, "Welfare State," 628.

27 The main asbestos mines are located in Canada, South Africa, Russia, and Brazil.
The most recent data on asbestos production and trade can be found at U.S.
Geological Survey, "Asbestos Statistics."

28 McCulloch and Tweedale, *Defending the Indefensible.*

29 Some decision-makers accepted the risks associated with asbestos exposure on
the grounds that asbestos increases safety of building construction even if less
dangerous substitutes were available early on. See Maines, *Asbestos and Fire.*

30 Stadler, "Corrosion Proof Fittings v. EPA."

31 McCulloch and Tweedale, *Defending the Indefensible.*

32 Peto et al., "Continuing Increase." In France, more than 1.2 million construction
workers face exposure risks; the same for 6 million workers in the United States.

33 In addition, occupational disease and death tends to be underrecorded. Rosental,
"Before Asbestos, Silicosis."

34 McCulloch and Tweedale, *Defending the Indefensible.*

35 Proctor, *Golden Holocaust*; Markowitz and Rosner, *Deceit and Denial.*

36 McCulloch and Tweedale, *Defending the Indefensible*, 119–154.

37 Castleman, "Controlled Use."

38 Thébaud-Mony, "Les fibres courtes."

39 Demortain, *Science of Bureaucracy.*

40 Paull, "Origin and Basis"; Murphy, *Sick Building Syndrome*; Reinhardt, "Limit Values"; Henry, "'License to Expose?'"

41 "The experts recommended that, in the present state of knowledge, the 2 fibres/
ml time-weighted standard which had been adopted by some member States
should be regarded as an interim target concentration for the prevention of risk to
the health of asbestos workers. It was recognised that this standard related to the
fibrogenic effects and not to the carcinogenic effects for which no standards exist
at the present time." International Labour Organization, "Report," 7.

42 Schwerin, "Low Dose Intoxication"; Schwerin, "Vom Gift im Essen"; Boudia,
"From Threshold to Risk."

43 Proctor, *Cancer Wars.*

44 Henry, *Ignorance scientifique*; Henry, "Governing Occupational Exposure."

45 Beck, *Risk Society*, 68.

46 Henry, *Amiante.*

47 An analogous case has to do with the low priority shown by the U.S. EPA in
regulating radon exposures in homes, despite the significant contribution of this
source of ionizing radiation to overall exposure in the general population. In general, the sources of radon are the natural environment. See U.S. Environmental
Protection Agency, *Unfinished Business.*

48 Counil, Daniau, and Isnard, "Étude de santé publique."

49 From 1938 to 1975, the factory's workforce ranged between ten and twenty-five
individuals. This figure almost certainly undercounts occupational disease since
many workers were immigrants who returned to their countries of origin or have
not claimed their rights to compensation.

50 Counil, "Place du chercheur."

51 Peto et al., "Continuing Increase"; Stayner, Welch, and Lemen, "Worldwide
Pandemic."

52 Since the 1990s, supply in different rare earths has become a critical political issue. From 2005, China had set up the export quotas of rare earths raw materials, which were then reduced by half between 2007 and 2010, from a maximum of sixty kilotons down to thirty kilotons of oxides of rare earths. In addition, in 2011 China raised export taxes on these materials from 15 percent to 25 percent.

53 Kiggins, *Political Economy*; Boudia, "Quand une crise."

54 *Le Monde*, "La Chine réduit"; Rijpma, "Terres rares." The idea is often attributed to President Deng Xiaoping, who is quoted as saying that "rare earths are to China what oil is to the Middle East."

55 Fear of disruption of rare earth supply has led to several public and private initiatives and reports. U.S. Department of Energy, *Critical Materials Strategy*, 2010; U.S. Department of Energy, *Critical Materials Strategy*, 2011; Ahonen et al., *Strengthening*.

56 Hopkins and Wallerstein defined a commodity chain as "a network of labour and production processes whose end result is a finished commodity." Hopkins and Wallerstein, "Commodity Chains," 159. Originated in the world-system school, the concept of the global commodity chain was later formulated as an analysis paradigm in a collected volume: Gereffi and Korzeniewicz, *Commodity Chains*.

57 Evans, *Episodes*; Abraham, "Rare Earths"; Klinger, "Historical Geography."

58 U.S. Geological Survey, *Mineral Commodity Summaries*, 2013.

59 Krishnamurthy and Gupta, *Extractive Metallurgy*; Balaram, "Rare Earth Elements."

60 Because their uses are so variable, there is no single market for rare earths, and consequently, production statistics remain difficult to track down and aggregate. Roskill Information Services, *Rare Earths*; Bureau de Recherches Géologiques et Minières, *Panorama*.

61 The Chinese production is relatively stable (120,000 tons), but its proportion in world production has fallen to 71 percent in 2018 and 62 percent in 2019 due to the revival of mining production in the United States and Australia. These data only partially reflect the complexity of the market, as part of the production outside China is carried out by Chinese companies. U.S. Geological Survey, *Mineral Commodity Summaries*, 2020.

62 Pagano et al., "Health Effects."

63 Reisman et al., *Rare Earth Elements*.

64 Maughan, "Dystopian Lake."

65 Drezet, "Les terres rares."

66 PR Newswire, "China Rare Earth."

67 Hurst, *China's Rare Earth*; Ali, "Social and Environmental"; Huang et al., "Protecting the Environment." This last study makes a state-of-the-art review of works on rare earth pollution, including an important set of works published in Chinese.

68 The city's official website burnishes this green image. See La Rochelle, "Sustainable City," June 18, 2021, https://www.larochelle.fr/action-municipale/ville-durable.

69 These wastes are known to contain thorium 232, uranium 238, and their decay products, notably radium 226 and 228.

70 The cross-referencing of various data from different records, official reports by French public authorities in charge of waste radioactive management and nuclear safety, the Agence nationale pour la gestion des déchets radioactifs (ANDRA), and the Autorité de sûreté nucléaire (ASN) allow us to trace part of the history

of this waste. In addition to several archives, the data provided in this paper are based on the following official reports: Agence nationale pour la gestion des déchets radioactifs, "Où sont les déchets radioactifs en france?"; Agence nationale pour la gestion des déchets radioactifs, "Les déchets à radioactivité naturelle renforcée (RNR)"; Autorité de sûreté nucléaire, *Bilan sur la gestion*; Autorité de sûreté nucléaire, *Plan national de gestion*.

71 The site is operated by Agence nationale pour la gestion des déchets radioactifs (ANDRA), the French nuclear waste management agency.

72 In French, "résidus solides banalisés," or RSB. See Agence nationale pour la gestion des déchets radioactifs, "Colis de Résidu Solide Banalisé."

73 In French, "matière en suspension." See Autorité de sûreté nucléaire, *Plan national de gestion*.

74 Hazardous waste is managed by the French Environment and Energy Management Agency (ADEME) and the Regional Environment, Planning and Housing Agency (DREAL); radioactive waste is managed by ANDRA.

75 Ministère de la Transition écologique et solidaire, France, "Pollution des sols."

76 United Nations Environment Programme, *Recycling Rates*.

77 In theory, the cost associated with rare earth recycling could become economically viable regarding Chinese restrictions on rare earths exports. However, charges from other countries of violating the General Agreement on Tariffs and Trade (GATT) and the World Health Organization's condemnation of China for this practice has led to an increase in the country's supply to the world market, hence to falling prices, which in turn make recycling uneconomical.

78 Pellow, *Resisting Global Toxics*; Little, *Burning Matters*.

Chapter 4: Apprehension

1 See, for instance, Dirzo et al., "Defaunation in the Anthropocene"; Habel et al., "Butterfly Community." See also Mikanowski, "Different Dimension."

2 Hallmann et al., "More Than 75 Percent."

3 Gabbatiss, "'Shocking' Decline."

4 Suryanarayanan and Kleinman, *Vanishing Bees*.

5 Mikanowski, "Different Dimension."

6 Foucart and Dagorn, "Pourquoi les pesticides."

7 On female labor, see Goulson, "Decline of Bees." On industry, see Suryanarayanan and Kleinman, *Vanishing Bees*. On pollinator drones, see Klein, "Robotic Bee"; and Potts et al., "Robotic Bees."

8 Kolbert, *Sixth Extinction*; Foucart, *Et le monde devint silencieux*.

9 Knowles, "Learning."

10 Our work builds on Shapiro, Zakariya, and Roberts, "Wary Alliance." See also Murphy on "regimes of perceptibility" and modes of apprehending chemical exposure in *Sick Building Syndrome*.

11 Giddens, *Consequences of Modernity*, 105 (on "ontological security," 92–100).

12 See, for instance, Auyero and Swistun, *Flammable*; Centemeri, "What Kind"; Davies, "Toxic Space and Time"; Shapiro, Zakariya, and Roberts, "Wary Alliance." See also Goldstein, "Invisible Harm"; and Calvillo et al., "Toxic Politics."

13 Office workers have had similar experiences: Murphy, *Sick Building Syndrome*.

14 Goldstein, "Invisible Harm," 321.

15 Alaimo, *Bodily Natures*, 4. For an insightful history that problematizes the sharp divide between organisms and their environment, see Benson, *Surroundings*.

16 Recent studies of the emergence of biomonitoring include Daemmrich, "Risk Frameworks"; Casper and Moore, "Fluid Matters"; and Morello-Frosch et al., "Toxic Ignorance."

17 Sexton, Needham, and Pirkle, "Human Biomonitoring."

18 van Sittert and de Jong, "Biomonitoring of Exposure." An alternative biomonitoring method involves searching for the *effects* of exposure through damage to genetic material (chromosomes or DNA) or other molecular alterations (usually called adducts) chemically induced by exposure. This has never become widely used at the populational level but has been used in occupational health surveillance. See Creager, "Human Bodies."

19 Centers for Disease Control and Prevention, *Second National Report*, 3.

20 Sexton, Needham, and Pirkle, "Human Biomonitoring," 40.

21 Sexton, Needham, and Pirkle, 21; Bertomeu-Sánchez, "Managing Uncertainty."

22 Kehoe, Thamann, and Cholak, "Lead Absorption"; Yant, Schrenk, and Patty, "Urine Sulfate Determinations."

23 Rosen, *History of Public Health*.

24 Markowitz and Rosner, *Lead Wars*.

25 Markowitz and Rosner, *Deceit and Denial*, 110.

26 Patterson, "Contaminated," 357.

27 For an important examination of this legacy, see Sellers, *Hazards of the Job*.

28 To be more specific, the CDC aims to study the health and nutrition status of a demographically representative group of noninstitutionalized civilians. Participants are selected at random and given comprehensive physical examinations, which include tissue samples for testing, and are also interviewed. National Research Council, *Monitoring Human Tissues*, 55.

29 Annest et al., "Chronological Trend." This publication made clear that there was a strong correlation between the decreased amount of lead used in the production of gasoline and blood lead levels in the United States. For a recent account, see Centers for Disease Control and Prevention, "National Health."

30 See Warren, *Brush with Death*, 224–243. In addition, by the late 1980s, clinicians had gathered evidence for damaging effects of low-level exposure to lead, especially in children. But industry fought hard against the public acceptance of research by Herbert Needleman linking low-level lead exposure to cognitive and behavioral deficits in children. See Markowitz and Rosner, *Lead Wars*, especially 87–121.

31 Evidence for this further drop came from the CDC's NHANES III survey, released in 1994. Pirkle et al., "Decline in Blood Lead Levels"; Warren, *Brush with Death*, 239.

32 National Research Council, *Monitoring Human Tissues*; U.S. General Accounting Office, *Toxic Chemicals*; National Research Council, *Human Biomonitoring*.

33 National Research Council, *Monitoring Human Tissues*, 2.

34 The number of chemicals identified in these tests depended and still depends on the limits of detection. In this sense, it may not reflect actual exposure.

35 U.S. General Accounting Office, *Toxic Chemicals*, 57. At the advice of the National Research Council, NHATS was discontinued in 1992 to be replaced by a more comprehensive national biomonitoring program. See National Research Council, *Monitoring Human Tissues*.

36 Before this time, NHANES included only a few heavy metals, pesticides, and volatile chemicals. In addition, in 1999, the CDC revised the design of NHANES so as to survey the same five thousand individuals annually. U.S. General Accounting

Office, *Toxic Chemicals*, 54–56. On the judgment about NHANES being the "first large biomonitoring study," see Szasz, *Shopping*, 100.

37 Centers for Disease Control and Prevention, *Second National Report*.
38 Schafer et al., *Chemical Trespass*.
39 Houlihan et al., *Body Burden*, 3.
40 Thornton, McCally, and Houlihan, "Biomonitoring."
41 World Wildlife Fund, "European Parliamentarians Contaminated."
42 World Wildlife Fund, "Bad Blood?"
43 Williams, Panko, and Paustenbach, "European Union's REACH Regulation."
44 National Research Council, *Human Biomonitoring*, 2.
45 National Research Council, 6.
46 Nancy Doerrer, scientific program manager at ILSI Health and Environmental Sciences Institute, an industry-funded outfit in Washington, D.C., as quoted in Stokstad, "Pollution Gets Personal," 1892.
47 American Chemistry Council, *Issue Brief*.
48 Liévanos et al., "Uneven Transformations."
49 Shostak, *Exposed Science*, 101–135.
50 Shapiro, Zakariya, and Roberts, "Wary Alliance," 577.
51 Brody et al., "Reporting Individual Results."
52 Altman et al., "Pollution Comes Home." The limitations on understanding what the biomonitoring exposure data means clinically is in part due to lack of resources but also scientific challenges in matching up epidemiological and monitoring data. Paustenbach and Galbraith, "Biomonitoring and Biomarkers"; Washburn, "Social Significance," 169–170.
53 Altman et al., "Pollution Comes Home," 421.
54 Shapiro, Zakariya, and Roberts, "Wary Alliance."
55 Alaimo, *Bodily Natures*, 108.
56 Silent Spring Institute, "Detox Me Action Kit."
57 Szasz, *Shopping*.
58 Boudia, "From Threshold to Risk"; Walker, *Permissible Dose*; Richards, "Rocks and Reactors."
59 Paull, "Origin and Basis"; Reinhardt, "Limit Values."
60 Sellers, *Hazards of the Job*.
61 Henry, "'License to Expose?'"
62 Castleman and Ziem, "Corporate Influence"; Roach and Rappaport, "But They Are Not."
63 Henry, *Ignorance scientifique*; Henry, "Governing Occupational Exposure."
64 Cordner, Richter, and Brown, "Chemical Class Approaches."
65 Robert, "Expert Groups."
66 Adam, "Industrial Food."
67 European Commission, "2014/113/EU."
68 Emmanuel Henry was the observer.
69 Hansson, *Setting the Limit*.
70 Käfferlein et al., "Human Exposure."
71 As Käfferlein et al. report,

> Our results on the controlled exposure to airborne aniline and the formation of Met-Hb at exposure levels of 2 ppm show a maximum increase of Met-Hb up to 2.1% and an increase of aniline in urine up to 420 μg/L. These levels are approximately twofold lower than the current German guidance levels for Met-Hb (5%, Leng and Bolt, 2008) and the biological threshold limit for aniline

in urine (1,000 μg/L, BMAS 2013) thus showing that the current threshold limits of aniline in air (2 ppm) and urine (1,000 μg/L) adequately protect workers from aniline-induced cyanosis via the formation of Met-Hb. Overall, our results contribute to the setting of exposure limits at the workplace. (1425)

72 Emmanuel Henry, Ethnographic Field Notebook (unpublished), 91st SCOEL meeting, March 12–13, 2014, Luxembourg.

73 Henry.

74 Lang, Bruckner, and Triebig, "Formaldehyde."

75 Mueller, Bruckner, and Triebig, "Exposure Study," 115.

76 "Forty-one male volunteers were exposed for 5 days (4 h per day) in a randomized schedule to the control condition (0 ppm) and to formaldehyde concentrations of 0.5 and 0.7 ppm and to 0.3 ppm with peak exposures of 0.6 ppm, and to 0.4 ppm with peak exposures of 0.8 ppm, respectively." Mueller, Bruckner, and Triebig, 107.

77 Henry, Ethnographic Field Notebook, 91st SCOEL meeting.

78 Beane Freeman et al., "Mortality from Solid Tumors."

79 Henry, Ethnographic Field Notebook (unpublished), 94th SCOEL meeting, December 3–4, 2014, Luxembourg.

80 Henry.

81 Bolt et al., "SCOEL/REC/125."

82 Gillam, *Whitewash*.

83 In 2010, Monsanto patented glyphosate as an antibiotic, showing its further versatility.

84 Monsanto was bought by Bayer in 2018.

85 Corporate Europe Observatory, "Glyphosate Saga."

86 Corporate Europe Observatory.

87 Majewski et al., "Pesticides in Mississippi."

88 Commissariat général au développement durable, "Les pesticides," 6.

89 Mills et al., "Excretion."

90 Sirinathsinghji and Ho, *Why Glyphosate*.

91 Corporate Europe Observatory, "European Citizens' Initiative."

92 The NGO Corporate Europe Observatory's regular syntheses have been especially useful to keep up with the myriad events occurring during those crucial years. See Glyphosate and Monsanto related articles on the Corporate Europe Observatory website (https://corporateeurope.org/).

93 For a detailed chronology of the 2014–2017 period, see Greenpeace European Unit, "EU Glyphosate Timeline."

94 Robinson, "Glyphosate Toxicity Studies."

95 Bauer-Panskus, *Testbiotech Comment*.

96 Neslen, "EU Report"; Corporate Europe Observatory, "Glyphosate Saga."

97 Guyton et al., "Carcinogenicity," 491.

98 In fact, Monsanto began planning its campaign against the IARC as soon as the agency announced that it would review glyphosate. They knew what would be the outcome—the classification as 2B or 2C—before the expert assessment took place. Under the California toxics law (passed in 1987 as Prop 65), IARC classifications of carcinogenicity are the basis for requiring labeling by chemical producers. The California move to regulate glyphosate became important globally because of how Monsanto's lawsuit against the state's action to regulate its product changed publicly available information about Monsanto. The court reviewing the case in California authorized the release of internal Monsanto documents

that had been submitted to it. European activists, among others, have used this information in their own work against glyphosate authorization.

99 Horel and Foucart, "Monsanto Papers, Part 1"; U.S. Right to Know, "Roundup (Glyphosate) Cancer Cases."
100 See, for instance, Piovano, "Project"; Baum Hedlund Aristei and Goldman PC, "Is Argentina Glyphosate"; and Rummel, *Tote Tiere, kranke Menschen.*
101 The European Parliament is an E.U. institution with limited power but a clear commitment to health and environmental issues. European Parliament, "Glyphosate Renewal."
102 Corporate Europe Observatory, "Glyphosate"; Corporate Europe Observatory, "Beneath."
103 See especially Arancibia, "Challenging the Bioeconomy"; Arancibia, "Regulatory Science"; and Arancibia et al., "Tensiones."
104 Arancibia, "Challenging the Bioeconomy," 84; Arancibia, "Regulatory Science"; Motta, *Social Mobilization.*
105 Arancibia, "Challenging the Bioeconomy," 85.
106 Arancibia, 86.
107 Arancibia, 86. See also Giarracca, Teubal, and Palmisano, "Paro agrario."
108 Arancibia, "Challenging the Bioeconomy," 86.
109 Arancibia, 86.
110 Arancibia, 87. Carrasco later published the study after peer review in the *International Journal of Toxicology.*
111 Arancibia, 86–87; Motta and Arancibia, "Health Experts."
112 See their website: http://reduas.com.ar/.
113 Koop, "Glyphosate Use"; Arancibia, "Regulatory Science."
114 Bandarage, "Political Economy"; de Silva, "Wanniyala-Aetto"; Gunawardene, "Science and Politics"; Elledge et al., "Chronic Kidney Disease."
115 Rajapakse, Shivanthan, and Selvarajah, "Chronic Kidney Disease."
116 Pearce and Caplin, "Let's Take the Heat Out."
117 Rajapakse, Shivanthan, and Selvarajah, "Chronic Kidney Disease"; Elledge et al., "Chronic Kidney Disease"; de Silva, "Bio-media Citizenship."
118 Jayasumana, Gunatilake, and Senanayake, "Glyphosate, Hard Water," 2125–2126. See also Jayasumana, Gunatilake, and Siribaddana, "Simultaneous Exposure."
119 Johnson et al., *Environmental Contamination.*
120 Bandarage, "Political Economy"; de Silva, "Wanniyala-Aetto"; de Silva, "Bio-media Citizenship."
121 Johnson et al., *Environmental Contamination.*
122 Bandarage, "Political Economy."
123 Glyphosate appears to have been prohibited first in 2011 along with a handful of other agrochemicals when a World Health Organization expert called for stronger regulations of "nephrotoxic agrochemicals." See Bandarage, 8; and Ranasinghe, "Chronic Kidney Disease."
124 de Silva, "Bio-media Citizenship," 8.
125 Wijedasa, "It's Official"; de Silva, "Govt. Lifts Glyphosate Ban."
126 de Silva, "Bio-media Citizenship."
127 de Silva.
128 Mason, "Mystery Kidney Disease."
129 de Silva, "Bio-media Citizenship," 8–11.
130 de Silva, 8.

131 Lock, *Encounters with Aging*; Lock, "Tempering." The quote is from Brotherton and Nguyen, "Revisiting Local Biology."

132 Nading, "Local Biologies," 142.

133 Hepler-Smith, "Molecular Bureaucracy."

134 Murphy, "Alterlife."

135 Benbrook, "How Did the US."

136 Demortain, *Science of Bureaucracy*.

137 Whether or not identification of residues rises to public acknowledgment as a problem or impels actual decisions are both empirical questions. We discuss the situation further in our conclusion.

138 See, for example, Michaels, *Doubt Is Their Product*; Oreskes and Conway, *Merchants of Doubt*; Proctor, *Golden Holocaust*.

Chapter 5: Residual Materialism

1 Bunker and Ciccantell, *Globalization*; Bunker, *Underdeveloping the Amazon*. Note that we do not use "residual materialism" here to refer to a remaining materialism in our thought.

2 Arcuri and Hendlin, "Chemical Anthropocene."

3 We owe this observation to Evan Hepler-Smith. Developed in the nineteenth century, coal gasification manufactured gas by heating coal in the absence of oxygen. The by-product of this process, called coal tar, in turn provided the base for the aniline dye industry. Cf. Travis, *Rainbow Makers*. On the link between coal and pollution, see Thorsheim, *Inventing Pollution*. On the origins and long-term environmental consequences of aniline dye chemistry, see Fagin, *Toms River*.

4 Ndiaye, *Nylon and Bombs*.

5 Colten, "Too Much"; Burby, "Baton Rouge."

6 Geiser, *Chemicals without Harm*.

7 Clapp, *Toxic Exports*; Alexander and Reno, *Economies of Recycling*; Lepawsky, *Reassembling Rubbish*.

8 We note the variety of numbers that get used in instances like this. 143,000 substances have been registered under the European regulatory system REACH. Yet a recent comprehensive analysis of 22 chemical inventories from 19 countries concludes that over 350,000 chemicals and mixtures of chemicals have been registered for production and use. It also shows that the identities of many chemicals remain publicly unknown. See Wang et al., "Toward a Global Understanding."

9 Homburg and Vaupel, *Hazardous Chemicals*, 2.

10 Brink et al., "Toward Sustainable Environmental Quality"; Furley et al., "Toward Sustainable Environmental Quality."

11 Markowitz and Rosner, *Deceit and Denial*; Oreskes and Conway, *Merchants of Doubt*; Michaels, *Triumph*; Appel, Mason, and Watts, *Subterranean Estates*; Davies, "Toxic Space and Time."

12 Henry, *Ignorance scientifique*; Boudia and Jas, *Gouverner*; Jas, "Millefeuilles institutionnels"; Hecht, "Work of Invisibility"; Bernardin and Henry, "Rationalization."

13 Hecht, "Work of Invisibility."

14 Cf. Latour, "Why Has Critique."

15 On environmental ignorance and agnotology, see Uekötter and Lübken, *Managing the Unknown*. More generally, see Proctor and Schiebinger, *Agnotology*; Gross and McGoey, *Routledge International Handbook*; McGoey, *Unknowers*; Girel, *Science et territoires*; and Richter et al., "Producing Ignorance."

16 Frickel, "On Missing New Orleans."

17 For an insightful recent overview of the Anthropocene debates, see Trischler, "Anthropocene"; Bonneuil and Fressoz, *Shock*; Chakrabarty, "Climate of History"; and Hecht, "Interscalar Vehicles."

18 McNeill and Engelke, *Great Acceleration*; Patel and Moore, *History of the World*; Chakrabarty, "Anthropocene Time"; Murphy, "Environment and Imperialism"; McNeill, Pádua, and Rangarajan, *Environmental History*; Arnold, *Toxic Histories*.

19 Vogel and Roberts, "Toxic Substances Control Act."

20 Dwiartama and Rosin, "Exploring Agency beyond Humans." See also Latour, *Politics of Nature*; Holifield, "Actor-Network Theory"; and Müller, "Assemblages and Actor-Networks."

21 U.S. Congress, House, *Waste Disposal Site Survey*; Dietrich, "Ultimate Disposal," 1–12.

22 Bonneuil and Fressoz, *Shock*, 81–82.

23 Murphy, *Economization of Life*, 137–138.

24 Jaramillo, "Mining Leftovers."

25 On the Capitalocene, see Moore, *Anthropocene or Capitalocene?* Interestingly, the debate is spawning other varieties of epochal world-making, such as the Plantationocene or the Chthulucene. See Davis et al., "Anthropocene"; and Haraway, *Staying*.

26 With this observation, we align with a number of writers who have reflected on the current environmental situation, including Haraway, *Staying*; Lynch and Veland, *Urgency*; Frickel, "Cities Are Subterranean Disasters"; Tsing, *Mushroom*; Fortun, "From Latour"; Liboiron, Tironi, and Calvillo, "Toxic Politics"; and Murphy, "Alterlife."

27 Haraway, *Staying*. See also Cohen and Galusky, "Guest Editorial."

28 Eyal, *Crisis*; Frickel and Arancibia, "Environmental STS"; Lidskog and Sundqvist, "Environmental Expertise"; McNeill, Pádua, and Rangarajan, *Environmental History*; McNeill and Engelke, *Great Acceleration*; Mitman, Armiero, and Emmett, *Future Remains*.

Bibliography

Abraham, Itty. "Rare Earths: The Cold War in the Annals of Travancore." In *Entangled Geographies: Empire and Technopolitics in the Global Cold War*, edited by Gabrielle Hecht, 101–124. Cambridge, Mass.: MIT Press, 2011.

Adam, Barbara. "Industrial Food for Thought: Timescapes of Risk." *Environmental Values* 8, no. 2 (1999): 219–238.

Adams, Robert. Interview by Lee Sullivan Berry. REACH Ambler, Chemical Heritage Foundation Oral History Collection, Science History Institute, November 22, 2013. Transcript #0811.

Agence nationale pour la gestion des déchets radioactifs. "Colis de Résidu Solide Banalisé RSB (SOLVAY), F6-8-02." Accessed July 3, 2020. https://inventaire.andra.fr/families/colis-de-residu-solide-banalise-rsb-solvay.

———. "Les déchets à radioactivité naturelle renforcée (RNR), Inventaire national des matières et déchets radioactifs." Chatenay-Malabry, France: Andra, 2012.

———. "Où sont les déchets radioactifs en France? Inventaire géographique des déchets radioactifs." Chatenay-Malabry, France: Andra, 2006.

Ahonen, Salla, Nikolaos Arvanitidis, Anton Auer, Emilie Baillet, Nazario Bellato, Koen Binnemans, Gian Andrea Blengini, Danilo Bonato, Ewa Brouwer, Sybolt Brower, et al. *Strengthening of the European Rare Earths Supply Chain: Challenges and Policy Options*. Brussels, Belgium: European Commission, 2015.

Aidala, James V. "The Toxic Substances Control Act: From the Perspective of James V. Aidala." Interview by Jody A. Roberts and Kavita D. Hardy. Chemical Heritage Foundation Oral History Collection, Science History Institute, May 20, 2010. Transcript #0660.

Alaimo, Stacy. *Bodily Natures: Science, Environment, and the Material Self*. Bloomington: Indiana University Press, 2010.

Alexander, Catherine, and Joshua Reno, eds. *Economies of Recycling: The Global Transformation of Materials, Values and Social Relations*. London: Zed, 2012.

Ali, Saleem H. "Social and Environmental Impact of the Rare Earth Industries." *Resources* 3, no. 1 (2014): 123–134.

Allen, Barbara L. *Uneasy Alchemy: Citizens and Experts in Louisiana's Chemical Corridor Disputes*. Cambridge, Mass.: MIT Press, 2003.

Altman, Rebecca Gasior, Rachel Morello-Frosch, Julia Green Brody, Ruthann Rudel, Phil Brown, and Mara Averick. "Pollution Comes Home and Gets Personal: Women's Experience of Household Chemical Exposure." *Journal of Health and Social Behavior* 49, no. 4 (2008): 417–435.

American Chemistry Council. *Issue Brief: Biomonitoring*. Washington, D.C.: American Chemistry Council, January 10, 2011. https://www.americanchemistry.com/Policy/Chemical-Safety/Biomonitoring/ACC-Issue-Brief-on-Biomonitoring.pdf.

Annest, Joseph L., James L. Pirkle, Diane Makuc, Jane W. Neese, David D. Bayse, and Mary Grace Kovar. "Chronological Trend in Blood Lead Levels between 1976 and 1980." *New England Journal of Medicine* 308, no. 23 (1983): 1373–1377. https://doi.org/10.1056/NEJM198306093082301.

Appel, Hannah, Arthur Mason, and Michael Watts, eds. *Subterranean Estates: Life Worlds of Oil and Gas*. Ithaca, N.Y.: Cornell University Press, 2015.

Arancibia, Florencia. "Challenging the Bioeconomy: The Dynamics of Collective Action in Argentina." *Technology in Society* 35, no. 2 (2013): 79–92. https://doi.org/DOI:10.3798/tia.1937-0237.16022.

———. "Regulatory Science and Social Movements: The Trial against the Use of Pesticides in Argentina." *Theory in Action* 9, no. 4 (2016): 1–21. https://doi.org/10.3798/tia.1937-0237.16022.

Arancibia, Florencia, Ignacio Bocles, Alicia Massarini, and Damián Verzeñassi. "Tensiones entre los saberes académicos y los movimientos sociales en las problemáticas ambientales." *Metatheoria* 8, no. 2 (2018): 105–123.

Arcuri, Alessandra, and Yogi Hale Hendlin. "The Chemical Anthropocene: Glyphosate as a Case Study of Pesticide Exposures." *King's Law Journal* 30, no. 2 (2019): 234–253.

Arnold, David. *Toxic Histories: Poison and Pollution in Modern India*. Cambridge: Cambridge University Press, 2016.

Autorité de sûreté nucléaire. *Bilan sur la gestion des déchets contenant de la radioactivité naturelle renforcée*. Montrouge, France: ASN, 2009.

———. *Plan national de gestion des matières et déchets radioactifs 2013–2015*. Montrouge, France: ASN, 2013.

Auyero, Javier, and Débora Alejandra Swistun. *Flammable: Environmental Suffering in an Argentine Shantytown*. Oxford: Oxford University Press, 2009.

Baccini, Peter, and Paul H. Brunner. *Metabolism of the Anthroposphere: Analysis, Evaluation, Design*. 2nd ed. Cambridge, Mass.: MIT Press, 2012.

Bachrach, Peter, and Morton S. Baratz. "Two Faces of Power." *American Political Science Review* 56, no. 4 (1962): 947–952.

Balaram, Vysetti. "Rare Earth Elements: A Review of Applications, Occurrence, Exploration, Analysis, Recycling, and Environmental Impact." *Geoscience Frontiers* 10, no. 4 (2019): 1285–1303.

Bandarage, Asoka. "Political Economy of Epidemic Kidney Disease in Sri Lanka." *SAGE Open* 3, no. 4 (2013): 1–13. https://doi.org/10.1177/2158244013511827.

Barad, Karen. *Meeting the Universe Halfway: Quantum Physics and the Entanglement of Matter and Meaning*. Durham, N.C.: Duke University Press, 2007.

Barles, Sabine. *L'invention des déchets urbains, France, 1790–1970*. Seyssel, France: Champ Vallon, 2005.

Bartrip, Peter. *Beyond the Factory Gates: Asbestos and Health in Twentieth Century America*. London: Continuum, 2006.

Bauer-Panskus, Andreas. *Testbiotech Comment on the German Renewal Assessment Report (RAR) on the Active Ingredient Glyphosate*. Munich: Testbiotech, 2014.

Baum Hedlund Aristei and Goldman PC. "Is Argentina Glyphosate Use Causing Health Issues in the Country?" December 11, 2017. https://www.baumhedlundlaw.com/12-17-argentina-glyphosate-health-issues/.

Beamish, Thomas D. *Silent Spill: The Organization of an Industrial Crisis.* Cambridge, Mass.: MIT Press, 2002.

Beane Freeman, Laura E., Aaron Blair, Jay H. Lubin, Patricia A. Stewart, Richard B. Hayes, Robert N. Hoover, and Michael Hauptmann. "Mortality from Solid Tumors among Workers in Formaldehyde Industries: An Update of the NCI Cohort." *American Journal of Industrial Medicine* 56, no. 9 (2013): 1015–1026.

Beck, Ulrich. *Risk Society: Towards a New Modernity.* London: Sage, 1992.

Beckert, Jens. *Imagined Futures: Fictional Expectations and Capitalist Dynamics.* Cambridge, Mass.: Harvard University Press, 2016.

Bell, Samuel, Thomas Marlow, Kai Wombacher, Anina Hitt, Neev Parikh, Andras Zsom, and Scott Frickel. "Automated Data Extraction from Historical City Directories: The Rise and Fall of Mid-century Gas Stations in Providence, RI." *PLoS ONE* 15, no. 8 (2020): e0220219. https://doi.org/10.1371/journal.pone.0220219.

Benbrook, Charles M. "How Did the US EPA and IARC Reach Diametrically Opposed Conclusions on the Genotoxicity of Glyphosate-Based Herbicides?" *Environmental Sciences Europe* 31, no. 2 (2019). https://doi.org/10.1186/s12302-018 -0184-7.

Bennett, Jane. *Vibrant Matter: A Political Ecology of Things.* Durham, N.C.: Duke University Press, 2010.

Benson, Etienne S. *Surroundings: A History of Environments and Environmentalism.* Chicago: University of Chicago Press, 2020.

Berenbaum, David, Dwyer Deighan, Thomas Marlow, Ashley Lee, Scott Frickel, and Mark Howison. "Mining Spatio-temporal Data on Industrialization from Historical Registries." *Journal of Environmental Informatics* 34, no. 1 (2019): 28–34. https://doi.org/10.3808/jei.201700381.

Bernadin, Stève, and Emmanuel Henry. "Rationalization, Privatization, Invisibilization? On Some Hidden Dimensions of the Transnationalization of Occupational Health and Traffic Safety Policies." In *Globalizing Issues: How Claims, Frames, and Problems Cross Borders,* edited by Erik Neveu and Muriel Surdez, 187–209. Cham, Switzerland: Palgrave Macmillan, 2020.

Bernal, J. D. *Science in History: The Natural Sciences in Our Time.* Vol. 3 of 4. Cambridge, Mass.: MIT Press, 1969.

Bertomeu-Sánchez, José Ramón. "Managing Uncertainty in the Academy and the Courtroom: Normal Arsenic and Nineteenth-Century Toxicology." *Isis* 104, no. 2 (2013): 197–225.

Biles, Blake A. "Harmonizing the Regulation of New Chemicals in the United States and in the European Economic Community." In *TSCA's Impact on Society and Chemical Industry,* edited by George W. Ingle, 39–65. ACS Symposium Series 213. Washington, D.C.: American Chemical Society, 1983. https://doi.org/10.1021/bk -1983-0213.ch004.

Boccuti, Salvatore A. Interview by Lee Sullivan Berry. REACH Ambler, Chemical Heritage Foundation Oral History Collection, Science History Institute, January 8, 2014. Transcript #0814.

Bohme, Susanna Rankin. *Toxic Injustice: A Transnational History of Exposure and Struggle.* Oakland: University of California Press, 2014.

Bolt, H. M., G. Johanson, G. D. Nielsen, D. Papameletiou, and C. L. Klein. *SCOEL/ REC/125 Formaldehyde: Recommendation from the Scientific Committee on Occupational Exposure Limits.* Brussels: European Commission, 2017. https://op.europa.eu/ en/publication-detail/-/publication/7a7aeoc9-co3d-11e6-a6db-01aa75ed71a1.

Bonneuil, Christophe, and Jean-Baptiste Fressoz. *The Shock of the Anthropocene: The Earth, History and Us*. Translated by David Fernbach. London: Verso, 2016.

Boudia, Soraya. "From Threshold to Risk: Exposure to Low Doses of Radiation and Its Effects on Toxicants Regulation." In *Toxicants, Health and Regulation since 1945*, edited by Soraya Boudia and Nathalie Jas, 71–87. London: Pickering & Chatto, 2013.

———. "Managing Scientific and Political Uncertainty: Environmental Risk Assessment in a Historical Perspective." In *Powerless Science? Science and Politics in a Toxic World*, edited by Soraya Boudia and Nathalie Jas, 95–112. New York: Berghahn, 2014.

———. "Quand une crise en cache une autre: La 'crise des terres rares' entre géopolitique, finance et dégâts environnementaux." *Critique internationale* 85, no. 4 (2019): 85–103.

Boudia, Soraya, Angela N. H. Creager, Scott Frickel, Emmanuel Henry, Nathalie Jas, Carsten Reinhardt, and Jody A. Roberts. "Residues: Rethinking Chemical Environment." *Engaging Science & Technology Studies* 4 (2018): 165–189.

Boudia, Soraya, and Nathalie Jas. *Gouverner un monde toxique*. Paris: Éditions Quae, 2019.

Boullier, Henri. "Évaluer des dossier 'vides'. L'expertise REACH face aux asymétries d'information." *Terrains & Travaux* 28, no. 1 (2016): 41–61.

———. *Toxiques légaux: Comment les firmes chimiques ont mis la main sur le contrôle de leurs produits*. Paris: La Découverte, 2019.

Boullier, Henri, David Demortain, and Maurice Zeeman. "Inventing Prediction for Regulation: The Development of (Quantitative) Structure-Activity Relationships for the Assessment of Chemicals at the US Environmental Protection Agency." *Science & Technology Studies* 32, no. 4 (2019): 137–157.

Bowker, Geoffrey C., and Susan Leigh Star. *Sorting Things Out: Classification and Its Consequences*. Cambridge, Mass.: MIT Press, 1999.

Boyland, E. "The Correlation of Experimental Carcinogenesis and Cancer in Man." *Progress in Experimental Tumor Research* 11 (1969): 222–234.

Brickman, Ronald, Sheila Jasanoff, and Thomas Ilgen. *Controlling Chemicals: The Politics of Regulation in Europe and the United States*. Ithaca, N.Y.: Cornell University Press, 1985.

Brink, Paul J. Van den, Alistair B. A. Boxall, Lorraine Maltby, Bryan W. Brooks, Murray A. Rudd, Thomas Backhaus, David Spurgeon, et al. "Toward Sustainable Environmental Quality: Priority Research Questions for Europe." *Environmental Toxicology and Chemistry* 37, no. 9 (2018): 2281–2295. https://doi.org/10.1002/etc .4205.

Brody, Julia Green, Sarah C. Dunagan, Rachel Morello-Frosch, Phil Brown, Sharyle Patton, and Ruthann A. Rudel. "Reporting Individual Results for Biomonitoring and Environmental Exposures: Lessons Learned from Environmental Communication Case Studies." *Environmental Health* 13, no. 40 (2014). https://doi.org/10 .1186/1476-069X-13-40.

Brotherton, P. Sean, and Vinh-Kim Nguyen. "Revisiting Local Biology in the Era of Global Health." *Medical Anthropology* 32, no. 4 (2013): 287–290. https://doi.org/10 .1080/01459740.2013.791290.

Brown, Kate. *Manual for Survival: A Chernobyl Guide to the Future*. New York: W. W. Norton, 2019.

Brown, Phil. *Toxic Exposures: Contested Illnesses and the Environmental Health Movement*. New York: Columbia University Press, 2007.

Bru, K., P. Christmann, J. F. Labbé, and G. Lefebvre. *Panorama 2014 du marché des Terres Rares.* Orléans, France: BRGM, 2015.

Buccini, John. *The Global Pursuit of the Sound Management of Chemicals.* Washington, D.C.: World Bank, 2004.

Buhs, Joshua Blu. *The Fire Ant Wars: Nature, Science, and Public Policy in Twentieth-Century America.* Chicago: University of Chicago Press, 2004.

Bullard, Robert D. *Dumping in Dixie: Race, Class, and Environmental Quality.* 3rd ed. Boulder, Colo.: Westview, 2000.

Bunker, Stephen G. "Modes of Extraction, Unequal Exchange, and the Progressive Underdevelopment of an Extreme Periphery: The Brazilian Amazon, 1600–1980." *American Journal of Sociology* 89, no. 5 (1984): 1017–1064. https://doi.org/10.1086/227983.

———. "Staples, Links, and Poles in the Construction of Regional Development Theories." *Sociological Forum* 4, no. 4 (1989): 589–610.

———. *Underdeveloping the Amazon: Extraction, Unequal Exchange, and the Failure of the Modern State.* Urbana: University of Illinois Press, 1985.

Bunker, Stephen G., and Paul S. Ciccantell. *Globalization and the Race for Resources.* Baltimore: Johns Hopkins University Press, 2005.

Burby, Raymond J. "Baton Rouge: The Making (and Breaking) of a Petrochemical Paradise." In *Transforming New Orleans & Its Environs: Centuries of Change,* edited by Craig E. Colten, 160–177. Pittsburgh, Pa.: University of Pittsburgh Press, 2000. https://doi.org/10.2307/j.ctt7zw9kz.18.

Bureau de Recherches Géologiques et Minières. *Panorama 2014 du Marché des Terres Rares.* Orléans, France: BRGM, 2015.

Burton, Ian, Robert W. Kates, and Gilbert F. White. *The Environment as Hazard.* Oxford: Oxford University Press, 1978.

Calvillo, Nerea, Max Liboiron, Manuel Tironi, and Nicole Nelson, eds. "Toxic Politics: Acting in a Permanently Polluted World." Special issue, *Social Studies of Science* 48, no. 3 (2018): 331–455.

Casper, Monica J., ed. *Synthetic Planet: Chemical Politics and the Hazards of Modern Life.* New York: Routledge, 2003.

Casper, Monica J., and Lisa Jean Moore. "Fluid Matters: Human Biomonitoring as Gendered Surveillance." In *Missing Bodies: The Politics of Visibility,* 109–131. New York: New York University Press, 2009.

Castleman, Barry. "'Controlled Use' of Asbestos." *International Journal of Occupational and Environmental Health* 9, no. 3 (2003): 294–298.

Castleman, Barry I., and Grace E. Ziem. "Corporate Influence on Threshold Limit Values." *American Journal of Industrial Medicine* 13, no. 5 (1988): 531–559.

Castleman, Barry I., and Stephen L. Berger. *Asbestos: Medical and Legal Aspects.* Englewood Cliffs, N.J.: Aspen, 2005.

Centemeri, Laura. "Investigating the 'Discrete Memory' of the Seveso Disaster in Italy." In *Governing Disasters: Beyond Risk Culture,* edited by Sandrine Revet and Julien Langumier, 191–219. New York: Palgrave Macmillan, 2015.

———. "What Kind of Knowledge Is Needed about Toxicant-Related Health Issues? Some Lessons Drawn from the Seveso Dioxin Case." In *Powerless Science? Science and Politics in a Toxic World,* edited by Soraya Boudia and Nathalie Jas, 134–151. New York: Berghahn, 2014.

Center of Excellence in Environmental Toxicology. "Mapping Ambler." REACH Ambler. Accessed June 21, 2021. https://ceet.upenn.edu/ambler_map/.

———. "From Factory to Future." REACH Ambler. Accessed June 21, 2021. https://ceet.upenn.edu/reachambler/.

Centers for Disease Control and Prevention. "National Health and Nutrition Examination Survey (NHANES) 2013–2014, Overview." 2014. https://wwwn.cdc.gov/
nchs/nhanes/ContinuousNhanes/Overview.aspx?BeginYear=2013.

———. *Second National Report on Human Exposure to Environmental Chemicals.*
Atlanta: U.S. Department of Health and Human Services, National Center for
Environmental Health, 2003.

Chakrabarty, Dipesh. "Anthropocene Time (The Seventh History and Theory Lec-
ture)." *History and Theory* 57, no. 1 (2018): 5–32.

———. "The Climate of History: Four Theses." *Critical Inquiry* 35, no. 2 (2009):
197–222.

Chen, Wan-ching G., and Thomas E. McKone. "Chronic Health Risks from
Aggregate Exposures to Ionizing Radiation and Chemicals: Scientific Basis for an
Assessment Framework." *Risk Analysis* 21, no. 1 (2001): 25–42.

Clapp, Jennifer. "Toxic Exports: Despite Global Treaty, Hazardous Waste Trade
Continues." In *Deviant Globalization: Black Market Economy in the 21st Century*,
edited by Nils Gilman, Jesse Goldhammer, and Steven Weber, 166–179. London:
Continuum, 2011.

———. *Toxic Exports: The Transfer of Hazardous Wastes from Rich to Poor Countries.*
Ithaca, N.Y.: Cornell University Press, 2001.

Clark, Claudia. *Radium Girls: Women and Industrial Health Reform, 1910–1935.* Chapel
Hill: University of North Carolina Press, 1997.

Cohen, Benjamin R., and Wyatt Galusky. "Guest Editorial." *Science as Culture* 19, no. 1
(2010): 1–14.

Cohen, Lizabeth. *A Consumer's Republic: The Politics of Mass Consumption in Postwar
America.* New York: Knopf, 2003.

Colborn, Theo, Dianne Dumanoski, and John Peterson Myers. *Our Stolen Future: Are
We Threatening Our Fertility, Intelligence, and Survival? A Scientific Detective Story.*
New York: Dutton, 1996.

Colten, Craig E. "Too Much of a Good Thing: Industrial Pollution in the Lower
Mississippi River." In *Transforming New Orleans & Its Environs*, edited by Craig E.
Colten, 141–159. Pittsburgh, Pa.: University of Pittsburgh Press, 2000. https://doi
.org/10.2307/j.ctt7zw9kz.17.

Commissariat général au développement durable. "Les pesticides dans les cours d'eau
français en 2013." *Chiffres & Statistiques*, no. 697 (2015). https://www.statistiques
.developpement-durable.gouv.fr/sites/default/files/2018-10/chiffres-stats697
-pesticides-dans-cours-deau2013-novembre2015.pdf.

"Comprehensive Environmental Response, Compensation, and Liability Act of 1980."
Code of Federal Regulations, 42 U.S.C., §§ 9601–9657 (2018).

Conner, Fred. Interview by Lee Sullivan Berry. REACH Ambler, Chemical Heritage
Foundation Oral History Collection, Science History Institute, September 14,
2014. Transcript #0917.

Cooke-Vargas, Sharon. Interview by Lee Sullivan Berry. REACH Ambler, Chemical
Heritage Foundation Oral History Collection, Science History Institute, September 15, 2014. Transcript #0918.

Cooper, Timothy. "Waste and 'Everyday Environmentalism' in Modern Britain." *Open
Library of Humanities* 3, no. 2 (2017): 3. https://doi.org/10.16995/olh.128.

Cordner, Alissa. *Toxic Safety: Flame Retardants, Chemical Controversies, and Environmental Health.* New York: Columbia University Press, 2016.

Cordner, Alissa, Lauren Richter, and Phil Brown. "Can Chemical Class Approaches
Replace Chemical-by-Chemical Strategies? Lessons from Recent U.S. FDA

Regulatory Action on Per- and Polyfluoroalkyl Substances." *Environmental Science & Technology* 50, no. 23 (2016): 12584–12591.

Corn, Jacqueline Karnell. *Protecting the Health of Workers: The American Conference of Governmental Industrial Hygienists, 1938–1988.* Cincinnati: American Conference of Governmental Industrial Hygienists, 1989.

Corporate Europe Observatory. "Beneath the Glyphosate Headlines, a Crucial Battle for the Future of EU Pesticide Approvals." October 24, 2017. https://corporateeurope.org/en/food-and-agriculture/2017/10/beneath-glyphosate -headlines-crucial-battle-future-eu-pesticide.

———. "European Citizens' Initiative to #StopGlyphosate: A Chance to Tackle Key Issues in Pesticide Safety and Agriculture?" February 8, 2017. https://corporateeurope.org/en/food-and-agriculture/2017/02/european-citizens-initiative -stopglyphosate-chance-tackle-key-issues.

———. "Glyphosate: One Pesticide, Many Problems." June 6, 2016. https://corporateeurope.org/en/food-and-agriculture/2016/06/glyphosate-one-pesticide -many-problems.

———. "The Glyphosate Saga, & 'Independent Scientific Advice' According to Germany, the UK & France." April 2, 2015. https://corporateeurope.org/en/food -and-agriculture/2015/04/glyphosate-saga-independent-scientific-advice-according -germany-uk.

Counil, Emilie. "La place du chercheur en épidémiologie dans la réponse à une alerte environnementale." *Environnement, Risques & Santé* 12, no. 4 (2013): 330–337.

Counil, Emilie, Côme Daniau, and Hubert Isnard. "Étude de santé publique autour d'une ancienne usine de broyage d'amiante: Le Comptoir des minéraux et matières premières à Aulnay-sous-Bois (Seine-Saint-Denis)—Pollution environnementale de 1938 et 1975; Impact sanitaire et recommandations." Saint-Maurice: Institut de veille sanitaire, 2007.

Cowles, Henry M. "Review of *Life Atomic: A History of Radioisotopes in Science and Medicine.*" *Medical History* 60, no. 4 (2016): 563–565.

———. *The Scientific Method: An Evolution of Thinking from Darwin to Dewey.* Cambridge, Mass.: Harvard University Press, 2020.

Cranor, Carl F. *Regulating Toxic Substances: A Philosophy of Science and the Law.* New York: Oxford University Press, 1993.

Creager, Angela N. H. "Human Bodies as Chemical Sensors: A History of Biomonitoring for Environmental Health and Regulation." *Studies in History and Philosophy of Science* 70 (2018): 70–81.

———. *Life Atomic: A History of Radioisotopes in Science and Medicine.* Chicago: University of Chicago Press, 2013.

———. "To Test or Not to Test: Tools, Rules, and Corporate Data in U.S. Chemicals Regulation." *Science, Technology, & Human Values* 46, no. 5 (2021): 975–997. https://doi.org/10.1177/01622439211013373.

Crooks, Ed. "BP Draws Line under Gulf Spill Costs." *Financial Times*, July 14, 2016. https://www.ft.com/content/ff2d8bcc-49e9-11e6-8d68-72e9211e86ab.

Crooks, Harold. *Giants of Garbage: The Rise of the Global Waste Industry and the Politics of Pollution Control.* Toronto: James Lorimer, 1993.

Crutzen, Paul J. "Geology of Mankind." *Nature* 415, no. 6867 (2002): 23.

Crutzen, Paul J., and Eugene F. Stoermer. "The 'Anthropocene.'" *IGBP Global Change Newsletter*, no. 41 (May 2000): 17–18.

Daemmrich, Arthur. "Risk Frameworks and Biomonitoring: Distributed Regulation of Synthetic Chemicals in Humans." *Environmental History* 13, no. 4 (2008): 684–694.

Dale, Amy L., Elizabeth A. Casman, Gregory V. Lowry, Jamie R. Lead, Enrica Viparelli, and Mohammed Baalousha. "Modeling Nanomaterial Environmental Fate in Aquatic Systems." *Environmental Science & Technology* 49, no. 5 (2015): 2587–2593.

Daniel, Pete. *Toxic Drift: Pesticides and Health in the Post-World War II South.* Baton Rouge: Louisiana State University Press, 2005.

Davies, J. Clarence. "The Toxic Substances Control Act: From the Perspective of J. Clarence Davies." Interview by Jody A. Roberts and Kavita D. Hardy. Chemical Heritage Foundation Oral History Collection. Science History Institute, October 30, 2009. Transcript #0640.

Davies, J. Clarence, and Barbara S. Davies. *The Politics of Pollution.* 2nd ed. Indianapolis: Pegasus, 1975.

Davies, J. Clarence, and Jan Mazurek. *Pollution Control in the United States: Evaluating the System.* Washington, D.C.: Resources for the Future, 1998.

Davies, Thom. "Toxic Space and Time: Slow Violence, Necropolitics, and Petrochemical Pollution." *Annals of the American Association of Geographers* 108, no. 6 (2018): 1537–1553.

Davis, Janae, Alex A. Moulton, Levi Van Sant, and Brian Williams. "Anthropocene, Capitalocene, . . . Plantationocene? A Manifesto for Ecological Justice in an Age of Global Crises." *Geography Compass* 13, no. 5 (2019): e12438.

Davis Le Blanc, Stacia. "Initiatives of Chemical Industry to Modify TSCA Regulations." In *TSCA's Impact on Society and Chemical Industry,* edited by George W. Ingle, 95–105. ACS Symposium Series 213. Washington, D.C.: American Chemical Society, 1983.

Davison, William. "Ethiopia's Deadly Rubbish Dump Landslide Was down to Politics, Not Providence." *Guardian,* March 24, 2017. https://www.theguardian.com/global-development/2017/mar/24/ethiopia-deadly-rubbish-dump-landslide-politics-not-providence-reppi.

Demortain, David. *The Science of Bureaucracy: Risk Decision-Making and the US Environmental Protection Agency.* Cambridge, Mass.: MIT Press, 2020.

Denison, Richard A. *Orphan Chemicals in the HPV Challenge: A Status Report.* New York: Environmental Defense Fund, 2004.

de Silva, Charumini. "Govt. Lifts Glyphosate Ban for Tea and Rubber." Daily Be Empowered, May 3, 2018. http://www.ft.lk/front-page/Govt--lifts-glyphosate-ban-for-tea-and-rubber/44-654413.

de Silva, M. W. Amarasiri. "Bio-media Citizenship and Chronic Kidney Disease of Unknown Etiology in Sri Lanka." *Medical Anthropology* 37, no. 3 (2018): 221–235. https://doi.org/10.1080/01459740.2017.1311886.

———. "The Wanniyala-Aetto, and CKDu." *Island Online,* November 14, 2017.

Dietrich, Gary N. "Ultimate Disposal of Hazardous Wastes." In *Impact of Legislation and Implementation on Disposal Management Practices,* edited by Robert B. Pojasek, 1–11. Vol. 3 of *Toxic and Hazardous Waste Disposal.* Ann Arbor: Ann Arbor Science, 1981.

Dijst, Martin, Ernst Worrell, Lars Böcker, Paul Brunner, Simin Davoudi, Stan Geertman, Robert Harmsen, et al. "Exploring Urban Metabolism—towards an Interdisciplinary Perspective." *Resources, Conservation & Recycling* 132 (2018): 190–203.

Dirzo, Rodolfo, Hillary S. Young, Mauro Galetti, Gerardo Ceballos, Nick J. B. Isaac, and Ben Collen. "Defaunation in the Anthropocene." *Science* 345, no. 6195 (2014): 401–406.

Douglas, Mary. *Purity and Danger: An Analysis of Concepts of Pollution and Taboo.* New York: Praeger, 1966.

Drezet, Eric. "Les terres rares: Quels impacts?" EcoInfo. August 6, 2010. https://ecoinfo.cnrs.fr/2010/08/06/les-terres-rares-quels-impacts/.

Dwiartama, Angga, and Christopher Rosin. "Exploring Agency beyond Humans: The Compatibility of Actor-Network Theory (ANT) and Resilience Thinking." *Ecology and Society* 19, no. 3 (2014). https://doi.org/10.5751/ES-06805-190328.

Egan, Michael. "Chronicling Quicksilver's Anthropogenic Cycle." *Global Environment* 7, no. 1 (2014): 10–37.

———. "Mercury's Web: Some Reflections on Following Nature across Time and Place." *Radical History Review*, no. 107 (Spring 2010): 111–126. https://doi.org/10.1215/01636545-2009-036.

———. "Toxic Knowledge: A Mercurial Fugue in Three Parts." *Environmental History* 13, no. 4 (2008): 636–642.

Elledge, Myles, Jennifer Hoponick Redmon, Keith Levine, R. J. Wickremasinghe, K. P. Wanigasariya, and R. J. Peiris-John. "Chronic Kidney Disease of Unknown Etiology in Sri Lanka: Quest for Understanding and Global Implications." In *RTI Press Research Brief*. Research Triangle Park, N.C.: RTI Press, 2014.

Environmental Defense Fund. *Toxic Ignorance: The Continuing Absence of Basic Health Testing for Top-Selling Chemicals in the United States.* New York: Environmental Defense Fund, 1997. http://catalog.hathitrust.org/api/volumes/oclc/38951043.html.

Erikson, Kai T. *A New Species of Trouble: The Human Experience of Modern Disasters.* New York: W. W. Norton, 1995.

Eriksson, Johan, Michael Gilek, and Christina Rudén, eds. *Regulating Chemical Risks: European and Global Challenges.* Dordrecht, Netherlands: Springer, 2010.

European Commission. "2014/113/EU; Commission Decision of 3 March 2014 on Setting Up a Scientific Committee on Occupational Exposure Limits for Chemical Agents and Repealing Decision 95/320/EC." *Official Journal of the European Union* L62 (March 4, 2014): 18–22.

European Food Safety Authority Panel on Contaminants in the Food Chain. "Presence of Microplastics and Nanoplastics in Food, with Particular Focus on Seafood." *EFSA Journal* 14, no. 6 (2016): e04501.

European Parliament. "Glyphosate Renewal: Parliament Puts People's Health First." News. November 28, 2017. https://www.europarl.europa.eu/news/en/headlines/society/20171124STO88814/glyphosate-renewal-parliament-puts-people-s-health-first.

Evans, C. H., ed. *Episodes from the History of the Rare Earth Elements.* Dordrecht, Netherlands: Kluwer Academic, 1996. https://doi.org/10.1007/978-94-009-0287-9.

Eyal, Gil. *The Crisis of Expertise.* Cambridge: Polity, 2019.

Fagin, Dan. *Toms River: A Story of Science and Salvation.* New York: Bantam, 2013.

Fellinger, Anne. "Du soupçon à la radioprotection: Les scientifiques face au risque professionnel de la radioactivité en France." PhD diss., University of Strasbourg, 2008.

Fortun, Kim. *Advocacy after Bhopal: Environmentalism, Disaster, New Global Orders.* Chicago: University of Chicago Press, 2001.

———. "From Latour to Late Industrialism." *HAU: Journal of Ethnographic Theory* 4, no. 1 (2014): 309–329.

Foster, John Bellamy. "Marx's Theory of Metabolic Rift: Classical Foundations for Environmental Sociology." *American Journal of Sociology* 105, no. 2 (1999): 366–405. https://doi.org/10.1086/210315.

Foucart, Stéphane. *Et le monde devint silencieux: Comment l'agrochimie a détruit les insectes.* Paris: Le Seuil, 2019.

Foucart, Stéphane, and Gary Dagorn. "Pourquoi les pesticides sont bien l'une des causes du déclin des oiseaux." *Le Monde*, March 29, 2018. https://www.lemonde .fr/les-decodeurs/article/2018/03/29/pourquoi-les-pesticides-sont-bien-l-une-des -causes-du-declin-des-oiseaux_5278294_4355770.html.

Frank, Arthur L., and T. K. Joshi. "The Global Spread of Asbestos." *Annals of Global Health* 80, no. 4 (2014): 257–262.

Frickel, Scott. *Chemical Consequences: Environmental Mutagens, Scientist Activism, and the Rise of Genetic Toxicology*. New Brunswick, N.J.: Rutgers University Press, 2004.

———. "Cities Are Subterranean Disasters." In *Feral Atlas: The More-Than-Human Anthropocene*, edited by Anna L. Tsing, Jennifer Deger, Alder Keleman Saxena, and Feifei Zhou. Stanford, Calif.: Stanford University Press Digital, 2020. https:// feralatlas.supdigital.org/poster/cities-are-subterranean-disasters.

———. "On Missing New Orleans: Lost Knowledge and Knowledge Gaps in an Urban Hazardscape." *Environmental History* 13, no. 4 (2008): 643–650.

Frickel, Scott, and Florencia Arancibia. "Environmental STS." In *International Handbook of Environmental Sociology*, edited by Beth Caniglia, Andrew Jorgenson, Stephanie Malin, Lori Peek, and David Pellow. London: Springer, 2021.

Frickel, Scott, and James R. Elliott. *Sites Unseen: Uncovering Hidden Hazards in American Cities*. New York: Russell Sage Foundation, 2018.

Frickel, Scott, Sahra Gibbon, Jeff Howard, Joanna Kempner, Gwen Ottinger, and David J. Hess. "Undone Science: Charting Social Movement and Civil Society Challenges to Research Agenda Setting." *Science, Technology, & Human Values* 35, no. 4 (2010): 444–473.

Frickel, Scott, and M. Bess Vincent. "Hurricane Katrina, Contamination, and the Unintended Organization of Ignorance." *Technology in Society* 29, no. 2 (2007): 181–188.

Furley, Tatiana Heid, Julie Brodeur, Helena C. Silva de Assis, Pedro Carriquiriborde, Katia R. Chagas, Jone Corrales, Marina Denadai, et al. "Toward Sustainable Environmental Quality: Identifying Priority Research Questions for Latin America." *Integrated Environmental Assessment and Management* 14, no. 3 (2018): 344–357.

Gabbatiss, John. "'Shocking' Decline in Birds across Europe Due to Pesticide Use, Say Scientists." *Independent*, March 21, 2018. https://www.independent.co.uk/ environment/europe-bird-population-countryside-reduced-pesticides-france -wildlife-cnrs-a8267246.html.

Gabrys, Jennifer, Gay Hawkins, and Mike Michael, eds. *Accumulation: The Material Politics of Plastic*. London: Routledge, 2013.

Gaynor, Kevin. "The Toxic Substances Control Act: A Regulatory Morass." *Vanderbilt Law Review* 30, no. 6 (1977): 1149–1196.

Geiser, Ken. *Chemicals without Harm: Policies for a Sustainable World*. Cambridge, Mass.: MIT Press, 2015.

Gereffi, Gary, and Miguel Korzeniewicz, eds. *Commodity Chains and Global Capitalism*. Westport, Conn.: Greenwood, 1994.

Geyer, Roland, Jenna R. Jambeck, and Kara Lavender Law. "Production, Use, and Fate of All Plastics Ever Made." *Science Advances* 3, no. 7 (2017): e1700782.

Giarracca, Norma, Miguel Teubal, and Tomás Palmisano. "Paro agrario: Crónica de un conflicto alargado." *Realidad económica* 237 (2008): 33–54.

Giddens, Anthony. *The Consequences of Modernity*. Stanford, Calif.: Stanford University Press, 1990.

Gillam, Carey. *Whitewash: The Story of a Weed Killer, Cancer, and the Corruption of Science*. Washington, D.C.: Island, 2017.

Girel, Mathias. *Science et territoires de l'ignorance*. Paris: Editions Quae, 2017.

Goldstein, Donna M., ed. "Invisible Harm: Science, Subjectivity and the Things We Cannot See." Special issue, *Culture, Theory and Critique* 58, no. 4 (2017): 321–456.

Goulson, Dave. "Decline of Bees Forces China's Apple Farmers to Pollinate by Hand." *China Dialogue* (blog), October 2, 2012. http://chinadialogue.net/en/uncategorized/5193-decline-of-bees-forces-china-s-apple-farmers-to-pollinate-by-hand/.

Gramaglia, Christelle. "Saltkrake: Penser la 'vitalité' des résidus miniers pour mieux appréhender leur effets toxiques." *Revue d'anthropologie des connaissances* 14, no. 4 (2020): 1–39. https://doi.org/10.4000/rac.11726.

Green Science Policy Institute. "Flame Retardants." *Green Science Policy Institute* (blog), October 14, 2013. https://greensciencepolicy.org/topics/flame-retardants/.

———. "September 2019: This Is Ridiculous!" *Green Science Policy Institute* (newsletter), September 2019. https://greensciencepolicy.org/news-events/newsletter/september-2019-this-is-ridiculous.

Greenpeace European Unit. "The EU Glyphosate Timeline." Greenpeace European Unit, August 2, 2017. https://www.greenpeace.org/eu-unit/issues/nature-food/1371/the-eu-glyphosate-timeline.

Gross, Matthias, and Linsey McGoey, eds. *Routledge International Handbook of Ignorance Studies*. New York: Routledge, 2015.

Gunawardene, Nalaka. "Science and Politics of Mass Kidney Failure in Sri Lanka." *Groundviews* (blog), August 19, 2012. https://groundviews.org/2012/08/19/science-and-politics-of-mass-kidney-failure-in-sri-lanka/.

Guyton, Kathryn Z., Dana Loomis, Yann Grosse, Fatiha El Ghissassi, Lamia Benbrahim-Tallaa, Neela Guha, Chiara Scoccianti, Heidi Mattock, Kurt Straif, and International Agency for Research on Cancer Monograph Working Group, IARC, Lyon, France. "Carcinogenicity of Tetrachlorvinphos, Parathion, Malathion, Diazinon, and Glyphosate." *Lancet* 16, no. 5 (2015): 490–491.

GZA Environmental, Inc. "Site Investigation Report: East Side Service Center." Providence, R.I., January 2006. Author files.

———. "Site Investigation Report—Addendum: East Side Service Center." Providence, R.I., November 2008. Author files.

Habel, Jan Christian, Andreas Segerer, Werner Ulrich, Olena Torchyk, Wolfgang W. Weisser, and Thomas Schmitt. "Butterfly Community Shifts over Two Centuries." *Conservation Biology* 30, no. 4 (2016): 754–762.

Hallmann, Caspar A., Martin Sorg, Eelke Jongejans, Henk Siepel, Nick Hofland, Heinz Schwan, Werner Stenmans, et al. "More Than 75 Percent Decline over 27 Years in Total Flying Insect Biomass in Protected Areas." *PLoS ONE* 12, no. 10 (2017): e0185809.

Hansen, Bjorn G., and Mark Blainey. "REACH: A Step Change in the Management of Chemicals." *Review of European Community & International Environmental Law* 15, no. 3 (2006): 270–280.

Hansson, Sven Ove. *Setting the Limit: Occupational Health Standards and the Limits of Science*. Oxford: Oxford University Press, 1998.

Haraway, Donna J. "A Cyborg Manifesto: Science, Technology, and Socialist-Feminism in the Late Twentieth Century." In *Simians, Cyborgs, and Women the Reinvention of Nature*, 149–181. New York: Routledge, 1991.

———. "Anthropocene, Capitalocene, Plantationocene, Chthulucene: Making Kin." *Environmental Humanities* 6, no. 1 (2015): 159–165.

———. *Staying with the Trouble: Making Kin in the Chthulucene*. Durham, N.C.: Duke University Press, 2016.

Harrison, Jill Lindsey. *Pesticide Drift and the Pursuit of Environmental Justice.* Cambridge, Mass.: MIT Press, 2011.

Harvie, David. *Deadly Sunshine: The History and Fatal Legacy of Radium.* Stroud, U.K.: Tempus, 2005.

Hecht, Gabrielle. *Being Nuclear: Africans and the Global Uranium Trade.* Cambridge, Mass.: MIT Press, 2014.

———. "Interscalar Vehicles for an African Anthropocene: On Waste, Temporality, and Violence." *Cultural Anthropology* 33, no. 1 (2018): 109–141.

———. "Residue." Somatosphere: Science, Medicine, and Anthropology. January 8, 2018. http://somatosphere.net/2018/residue.html/.

———. "The Work of Invisibility: Radiation Hazards and Occupational Health in South African Uranium Production." *International Labor and Working-Class History*, no. 81 (2012): 94–113.

Henry, Emmanuel. *Amiante, un scandale improbable: Sociologie d'un problème public.* Rennes, France: Presses Universitaires de Rennes, 2007.

———. "Governing Occupational Exposure Using Thresholds: A Policy Biased Toward Industry." *Science, Technology, & Human Values* 46, no. 5 (2021): 953–974. https://doi.org/10.1177/0162243921101530.

———. *Ignorance scientifique et inaction publique: Les politiques de santé au travail.* Paris: Presses de Sciences Po, 2017.

———. "'License to Expose'? Occupational Exposure Limits, Scientific Expertise and State in Contemporary France." In *Toxicants, Health and Regulation since 1945*, edited by Nathalie Jas and Soraya Boudia, 89–102. London: Pickering & Chatto, 2013.

Hepler-Smith, Evan. "Molecular Bureaucracy: Toxicological Information and Environmental Protection." *Environmental History* 24, no. 3 (2019): 534–560.

Hess, David J. "The Politics of Niche-Regime Conflicts: Distributed Solar Energy in the United States." *Environmental Innovation and Society Transitions* 19 (2016): 42–50.

Heyvaert, Veerle. "No Data, No Market: The Future of EU Chemicals Control under the REACH Regulation." *Environmental Law Review* 9, no. 3 (2007): 201–206.

———. "Reconceptualizing Risk Assessment." *Review of European Community & International Environmental Law* 8, no. 2 (1999): 135–143.

Higginson, J. "Present Trends in Cancer Epidemiology." *Proceedings of the Canadian Cancer Conference* 8 (1969): 40–75.

Hobsbawm, Eric J. "Introduction: Inventing Traditions." In *The Invention of Tradition*, edited by Eric J. Hobsbawm and Terrence Ranger, 1–14. Cambridge: Cambridge University Press, 1983.

———. "The Social Function of the Past: Some Questions." *Past & Present* 55 (1972): 3–17.

Holifield, Ryan. "Actor-Network Theory as a Critical Approach to Environmental Justice: A Case against Synthesis with Urban Political Ecology." *Antipode* 41, no. 4 (2009): 637–658. https://doi.org/10.1111/j.1467-8330.2009.00692.x.

Homburg, Ernst, and Elisabeth Vaupel, eds. *Hazardous Chemicals: Agents of Risk and Change, 1800–2000.* New York: Berghahn, 2019.

Economist. "Home on a Mountain of Rubbish." July 13, 2000. https://www.economist.com/asia/2000/07/13/home-on-a-mountain-of-rubbish.

Hooks, Gregory, and Chad L. Smith. "The Treadmill of Destruction: National Sacrifice Areas and Native Americans." *American Sociological Review* 69, no. 4 (2004): 558–575.

Hoover, Elizabeth. *The River Is in Us: Fighting Toxics in a Mohawk Community*. Minneapolis: University of Minnesota Press, 2017.

Hopkins, Terence K., and Immanuel Wallerstein. "Commodity Chains in the World-Economy Prior to 1800." *Review (Fernand Braudel Center)* 10, no. 1 (1986): 157–170.

Horel, Stéphane, and Stéphane Foucart. "The Monsanto Papers, Part 1—Operation: Intoxication." Environmental Health News, November 20, 2017. https://www.ehn.org/monsanto-glyphosate-cancer-smear-campaign-2509710888.html.

Houlihan, Jane, Richard Wiles, Kris Thayer, and Sean Gray. *Body Burden: The Pollution in People*. Washington, D.C.: Environmental Working Group, 2003.

Huang, Xiang, Guochun Zhang, An Pan, Fengying Chen, and Chunli Zheng. "Protecting the Environment and Public Health from Rare Earth Mining." *Earth's Future* 4, no. 11 (2016): 532–535.

Hurst, Cindy. *China's Rare Earth Elements Industry: What Can the West Learn?* Washington, D.C.: Institute for the Analysis of Global Security, 2010.

Ilgen, Thomas L. "'Better Living through Chemistry': The Chemical Industry in the World Economy." *International Organization* 37, no. 4 (1983): 647–680.

Ingle, George W. "Background, Goals, and Resultant Issues." In *TSCA's Impact on Society and Chemical Industry*, edited by George W. Ingle, 1–6. ACS Symposium Series 213. Washington, D.C.: American Chemical Society, 1983. https://doi.org/10.1021/bk-1983-0213.ch001.

International Labour Organization. "Report of the Meeting of Experts on the Safe Use of Asbestos, Geneva, 11–18 Dec 1973." Geneva: ILO, 1974.

Interpol. "Hazardous Materials Seized in Largest Global Operation against Illegal Waste." August 8, 2017. https://www.interpol.int/en/News-and-Events/News/2017/Hazardous-materials-seized-in-largest-global-operation-against-illegal-waste.

Jalbert, Kirk, Anna J. Willow, David Casagrande, and Stephanie Paladino, eds. *ExtrACTION: Impacts, Engagements, and Alternative Futures*. London: Routledge, 2017.

Jambeck, Jenna R., Roland Geyer, Chris Wilcox, Theodore R. Siegler, Miriam Perryman, Anthony Andrady, Ramani Narayan, and Kara Lavender Law. "Plastic Waste Inputs from Land into the Ocean." *Science* 347, no. 6223 (2015): 768–771.

Jaramillo, Pablo. "Mining Leftovers: Making Futures on the Margins of Capitalism." *Cultural Anthropology* 35, no. 1 (2020): 48–73. https://doi.org/10.14506/ca35.1.07.

Jarrige, François, and François Le Roux. *The Contamination of the Earth: A History of Pollutions in the Industrial Age*. Cambridge, Mass.: MIT Press, 2020.

Jas, Nathalie. "Adapting to 'Reality': The Emergence of International Expertise on Food Additives and Contaminants in the 1950s and Early 1960s." In *Toxicants, Health and Regulation since 1945*, edited by Soraya Boudia and Nathalie Jas, 47–70. London: Pickering & Chatto, 2013.

———. "Gouverner les substances chimiques dangereuses dans les espaces internationaux." In *Le gouvernement des technosciences: Gouverner le progrès et ses dégâts depuis 1945*, edited by Dominique Pestre, 31–63. Paris: La Découverte, 2014.

———. "Millefeuilles institutionnels et production d'ignorance dans le 'gouvernement' des substances chimiques dangereuses." *Raison présente* 204 (2017): 43–52.

Jasanoff, Sheila. *States of Knowledge: The Co-production of Science and Social Order*. London: Routledge, 2004.

———. *The Fifth Branch: Science Advisers as Policymakers*. Cambridge, Mass.: Harvard University Press, 1990.

Jasanoff, Sheila, and Dogan Perese. "Welfare State or Welfare Court: Asbestos Litigation in Comparative Perspective." *Journal of Law and Policy* 12, no. 2 (2004): 619–639.

Jayasumana, Channa, Sarath Gunatilake, and Priyantha Senanayake. "Glyphosate, Hard Water and Nephrotoxic Metals: Are They the Culprits behind the Epidemic of Chronic Kidney Disease of Unknown Etiology in Sri Lanka?" *International Journal of Environmental Research and Public Health* 11, no. 2 (2014): 2125–2147.

Jayasumana, Channa, Sarath Gunatilake, and Sisira Siribaddana. "Simultaneous Exposure to Multiple Heavy Metals and Glyphosate May Contribute to Sri Lankan Agricultural Nephropathy." *BMC Nephrology* 16, no. 1 (2015): 103.

Johnson, Sapna, Savvy Soumya Misra, Ramakant Sahu, and Poornima Saxena. *Environmental Contamination and Its Association with Chronic Kidney, Disease of Unknown Etiology in North Central Region of Sri Lanka.* New Delhi: Centre for Science and Environment, 2012.

Käfferlein, Heiko Udo, Horst Christoph Broding, Jürgen Bünger, Birger Jettkant, Stephan Koslitz, Martin Lehnert, Eike Maximilian Marek, et al. "Human Exposure to Airborne Aniline and Formation of Methemoglobin: A Contribution to Occupational Exposure Limits." *Archives of Toxicology* 88, no. 7 (2014): 1419–1426.

Karuga, James. "Largest Landfills, Waste Sites, and Trash Dumps in the World." WorldAtlas. Accessed June 27, 2020. https://www.worldatlas.com/articles/largest -landfills-waste-sites-and-trash-dumps-in-the-world.html.

Kehoe, Robert A., Frederick Thamann, and Jacob Cholak. "Lead Absorption and Excretion in Certain Lead Trades." *Journal of Industrial Hygiene* 15 (1933): 306–319.

Kellow, Aynsley J. *International Toxic Risk Management: Ideals, Interests and Implementation.* Cambridge: Cambridge University Press, 1999.

Khetan, Sushil K. *Endocrine Disruptors in the Environment.* Hoboken, N.J.: John Wiley & Sons, 2014.

Kiggins, Ryan David. *The Political Economy of Rare Earth Elements: Rising Powers and Technological Change.* New York: Palgrave Macmillan, 2015.

Klein, Alice. "Robotic Bee Could Help Pollinate Crops as Real Bees Decline." *New Scientist*, February 9, 2017. https://www.newscientist.com/article/2120832-robotic -bee-could-help-pollinate-crops-as-real-bees-decline/.

Klinger, Julie Michelle. "A Historical Geography of Rare Earth Elements: From Discovery to the Atomic Age." *Extractive Industries and Society* 2, no. 3 (2015): 572–580. https://doi.org/10.1016/j.exis.2015.05.006.

Knowles, Scott Gabriel. "Learning from Disaster? The History of Technology and the Future of Disaster Research." *Technology and Culture* 55, no. 4 (2014): 773–784.

Kolbert, Elizabeth. *The Sixth Extinction: An Unnatural History.* New York: Henry Holt, 2014.

Koop, Fermín. "Glyphosate Use on the Rise in Argentina, despite Controversy." *Buenos Aires Times*, January 13, 2018. https://www.batimes.com.ar/news/economy/ glyphosate-use-on-the-rise-in-argentina-despite-controversy.phtml.

Koselleck, Reinhart. *Futures Past: On the Semantics of Historical Time.* Translated by Keith Tribe. New York: Columbia University Press, 2004.

———. *Vergangene Zukunft: Zur Semantik geschichtlicher Zeiten.* Frankfurt: Suhrkamp Verlag, 1979.

Krimsky, Sheldon. *Hormonal Chaos: The Scientific and Social Origins of the Environmental Endocrine Hypothesis.* Baltimore: Johns Hopkins University Press, 2000.

Krishnamurthy, Nagaiyar, and Chiranjib Kumar Gupta. *Extractive Metallurgy of Rare Earths.* 2nd ed. Boca Raton, Fla.: CRC, 2015.

La Rochelle. "Sustainable City." June 18, 2021. https://www.larochelle.fr/action -municipale/ville-durable.

Landecker, Hannah. "Antibiotic Resistance and the Biology of History." *Body & Society* 22, no. 4 (2016): 19–52.

Lang, Isabelle, Thomas Bruckner, and Gerhard Triebig. "Formaldehyde and Chemosensory Irritation in Humans: A Controlled Human Exposure Study." *Regulatory Toxicology and Pharmacology* 50, no. 1 (2008): 23–36.

Langston, Nancy. *Toxic Bodies: Hormone Disruptors and the Legacy of DES.* New Haven, Conn.: Yale University Press, 2010.

Lanier-Christensen, Colleen. "Creating Regulatory Harmony: The Participatory Politics of OECD Chemical Testing Standards in the Making," *Science, Technology, & Human Values* 46, no. 5 (2021): 925–952. https://doi.org/10.1177/0162243921102936o.

Latour, Bruno. *Politics of Nature: How to Bring the Sciences into Democracy.* Translated by Catherine Porter. Cambridge, Mass.: Harvard University Press, 2009.

———. *Reassembling the Social: An Introduction to Actor-Network-Theory.* Oxford: Oxford University Press, 2005.

———. "Why Has Critique Run Out of Steam? From Matters of Fact to Matters of Concern." *Critical Inquiry* 30, no. 2 (2004): 225–248.

Law, John. "The Materials of STS." In *The Oxford Handbook of Material Culture Studies,* edited by Dan Hicks and Mary C. Beaudry, 171–186. Oxford: Oxford University Press, 2010.

Le Monde. "La Chine réduit ses exportations de terres rares pour début 2011." December 28, 2010. https://www.lemonde.fr/planete/article/2010/12/28/la-chine-reduit-ses -exportations-de-terres-rares-pour-debut-2011_1458556_3244.html.

Lepawsky, Josh. *Reassembling Rubbish: Worlding Electronic Waste.* Cambridge, Mass.: MIT Press, 2018.

Liboiron, Max. "Redefining Pollution and Action: The Matter of Plastics." *Journal of Material Culture* 21, no. 1 (2016): 87–110.

Liboiron, Max, Manuel Tironi, and Nerea Calvillo. "Toxic Politics: Acting in a Permanently Polluted World." *Social Studies of Science* 48, no. 3 (2018): 331–349.

Library of Congress. *Legislative History of the Toxic Substances Control Act.* Washington, D.C.: U.S. Government Printing Office, 1976.

Lidskog, Rolf, and Göran Sundqvist. "Environmental Expertise as Group Belonging: Environmental Sociology Meets Science and Technology Studies." *Nature and Culture* 13, no. 3 (2018): 309–331.

Liévanos, Raoul S., Jonathan K. London, Julie Sze, and Kim Fortun. "Uneven Transformations and Environmental Justice: Regulatory Science, Street Science, and Pesticide Regulation in California." In *Technoscience and Environmental Justice,* edited by Gwen Ottinger and Benjamin Cohen, 201–228. Cambridge, Mass.: MIT Press, 2011. https://doi.org/10.7551/mitpress/9780262015790.003.0009.

Little, Peter C. *Burning Matters: Life, Labor, and E-waste Pyropolitics in Ghana.* New York: Oxford University Press, forthcoming.

Lock, Margaret. *Encounters with Aging: Mythologies of Menopause in Japan and North America.* Berkeley: University of California Press, 1993.

———. "The Tempering of Medical Anthropology: Troubling Natural Categories." *Medical Anthropology Quarterly* 15, no. 4 (2001): 478–492.

Long, Bill L. *International Environmental Issues and the OECD, 1950–2000: A Historical Perspective.* Paris: OECD, 2000.

Lönngren, Rune. *International Approaches to Chemicals Control: A Historical Overview.* Solna, Sweden: Kemikalieinspektionen (KEMI), the National Chemicals Inspectorate, 1992.

Luckin, Bill. "Pollution in the City." In *The Cambridge Urban History of Britain: Volume 3: 1840–1950*, edited by Martin Daunton, 3:207–228. Cambridge: Cambridge University Press, 2001. https://doi.org/10.1017/CHOL9780521417075.008.

Lynch, Amanda H., and Siri Veland. *Urgency in the Anthropocene*. Cambridge, Mass.: MIT Press, 2018.

Maines, Rachel. *Asbestos and Fire: Technological Tradeoffs and the Body at Risk*. New Brunswick, N.J.: Rutgers University Press, 2005.

Majewski, Michael S., Richard H. Coupe, William T. Foreman, and Paul D. Capel. "Pesticides in Mississippi Air and Rain: A Comparison between 1995 and 2007." *Environmental Toxicology and Chemistry* 33, no. 6 (2014): 1283–1293. https://doi.org/10.1002/etc.2550.

Maloney, J. P. Sean. "A Legislative History of Liability under CERCLA Superfund Liability." *Seton Hall Legislative Journal* 16, no. 2 (1992): 517–550.

Markowitz, Gerald E., and David Rosner. *Deceit and Denial: The Deadly Politics of Industrial Pollution*. Berkeley: University of California Press, 2002.

———. *Lead Wars: The Politics of Science and the Fate of America's Children*. Berkeley: University of California Press, 2013.

Marlow, Thomas, Scott Frickel, and James R. Elliott. "Do Legacy Industrial Sites Produce Legacy Effects in Ethnic and Racial Residential Settlement? Environmental Inequality Formation in Rhode Island's Industrial Core." *Sociological Forum* 35, no. 4 (2021): 1093–1113.

Martin, Emily. "Anthropology and the Cultural Study of Science." *Science, Technology, & Human Values* 23, no. 1 (1998): 24–44.

Mason, Margie. "Mystery Kidney Disease Killing Sri Lankan Farmers." *AP News*, January 18, 2015. https://apnews.com/0e81050a0d1340e6ac2727f5ef762e09.

Maughan, Tim. "The Dystopian Lake Filled by the World's Tech Lust." *BBC*, April 2, 2015. http://www.bbc.com/future/story/20150402-the-worst-place-on-earth.

McCulloch, Jock. *Asbestos Blues: Labour, Capital, Physicians & the State in South Africa*. Bloomington: Indiana University Press, 2002.

McCulloch, Jock, and Geoffrey Tweedale. *Defending the Indefensible: The Global Asbestos Industry and Its Fight for Survival*. Oxford: Oxford University Press, 2008.

McDonough, Anne. Interview by Lee Sullivan Berry. REACH Ambler, Chemical Heritage Foundation Oral History Collection, Science History Institute, January 20, 2014. Transcript #0933.

McGoey, Linsey. *The Unknowers: How Strategic Ignorance Rules the World*. London: Zed, 2019.

McMahon, Tim. "Historical Crude Oil Prices (Table)." InflationData.com, May 21, 2020. https://inflationdata.com/Inflation/Inflation_Rate/Historical_Oil_Prices_Table.asp.

McNeill, John Robert, and Peter Engelke. *The Great Acceleration: An Environmental History of the Anthropocene since 1945*. Cambridge, Mass.: Belknap, 2014.

McNeill, John Robert, José Augusto Pádua, and Mahesh Rangarajan, eds. *Environmental History: As If Nature Existed*. Oxford: Oxford University Press, 2010.

McNeill, John Robert, and George Vrtis, eds. *Mining North America: An Environmental History since 1522*. Oakland: University of California Press, 2017.

Melosi, Martin V. *Garbage in the Cities: Refuse, Reform, and the Environment*. Revised edition. Pittsburgh, Pa.: University of Pittsburgh Press, 2005.

———. *The Sanitary City: Urban Infrastructure in America from Colonial Times to the Present*. Baltimore: Johns Hopkins University Press, 2000.

Michaels, David. *Doubt Is Their Product: How Industry's Assault on Science Threatens Your Health*. New York: Oxford University Press, 2007.

———. *The Triumph of Doubt: Dark Money and the Science of Deception*. Oxford: Oxford University Press, 2020.

Mikanowski, Jacob. "'A Different Dimension of Loss': Inside the Great Insect Die-Off." *Guardian*, December 14, 2017. https://www.theguardian.com/environment/2017/dec/14/a-different-dimension-of-loss-great-insect-die-off-sixth-extinction.

Mills, Paul J., Izabela Kania-Korwel, John Fagan, Linda K. McEvoy, Gail A. Laughlin, and Elizabeth Barrett-Connor. "Excretion of the Herbicide Glyphosate in Older Adults between 1993 and 2016." *Journal of the American Medical Association* 318, no. 16 (2017): 1610–1611.

Ministère de la Transition Écologique et Solidaire, France. "Pollution des sols: BASOL." Accessed June 21, 2021. https://www.georisques.gouv.fr/risques/sites-et-sols-pollues/donnees?page=1&index_sp=17.0011#/.

Mitman, Gregg, Marco Armiero, and Robert S. Emmett, eds. *Future Remains: A Cabinet of Curiosities for the Anthropocene*. Chicago: University of Chicago Press, 2018.

Moll-François, Fabien. "Problématiser les contaminations, mettre en cause les responsables: Mobilisations, expertises et recours au droit pénal dans les affaires amiante et dioxines en France (1975–2015)." PhD diss., École des hautes études en sciences sociales, 2012.

Monsaingeon, Baptiste. *Homo detritus: Critique de la société du déchet*. Paris: Éditions du Seuil, 2017.

Moore, Amelia, ed. "The Anthropocene: A Critical Exploration." Special issue. *Environment and Society* 6, no. 1 (2016): 1–166.

Moore, Charles. "Trashed: Across the Pacific Ocean, Plastics, Plastics, Everywhere." *Natural History*, November 2003. https://www.naturalhistorymag.com/htmlsite/1103/1103_feature.html.

Moore, Jason W., ed. *Anthropocene or Capitalocene? Nature, History, and the Crisis of Capitalism*. Oakland, Calif.: PM Press, 2016.

Morello-Frosch, Rachel, Julia Green Brody, Phil Brown, Rebecca Gasior Altman, Ruthann A. Rudel, and Carla Pérez. "Toxic Ignorance and Right-to-Know in Biomonitoring Results Communication: A Survey of Scientists and Study Participants." *Environmental Health* 8, no. 1 (2009): 6.

Mosbergen, Dominique. "New Plastic Garbage Patch Found in the South Pacific Could Be '1.5 Times Larger Than Texas.'" *Huffington Post*, August 2, 2017. https://www.huffingtonpost.ca/entry/south-pacific-garbage-patch_n_59818f92e4b0353fbb3387ac?fz=&utm_hp_ref=ca-world.

Motta, Renata. *Social Mobilization, Global Capitalism and Struggles over Food: A Comparative Study of Social Movements*. London: Routledge, 2016.

Motta, Renata, and Florencia Arancibia. "Health Experts Challenge the Safety of Pesticides in Argentina and Brazil." In *Medicine, Risk, Discourse and Power*, edited by John Martyn Chamberlain, 179–206. New York: Routledge, 2016.

Mueller, Joerg U., Thomas Bruckner, and Gerhard Triebig. "Exposure Study to Examine Chemosensory Effects of Formaldehyde on Hyposensitive and Hypersensitive Males." *International Archives of Occupational and Environmental Health* 86, no. 1 (2013): 107–117.

Müller, Martin. "Assemblages and Actor-Networks: Rethinking Socio-Material Power, Politics and Space." *Geography Compass* 9, no. 1 (2015): 27–41. https://doi.org/10.1111/gec3.12192.

Murphy, Joseph. "Environment and Imperialism: Why Colonialism Still Matters." SRI Papers (Sustainability Research Institute), no. 20, School of Earth and Environment, University of Leeds, U.K., October 2009. http://www.see.leeds.ac.uk/fileadmin/Documents/research/sri/workingpapers/SRIPs-20_02.pdf.

Murphy, Michelle. "Alterlife and Decolonial Chemical Relations." *Cultural Anthropology* 32, no. 4 (2017): 494–503.

———. *The Economization of Life*. Durham, N.C.: Duke University Press, 2017.

———. *Sick Building Syndrome and the Problem of Uncertainty: Environmental Politics, Technoscience, and Women Workers*. Durham, N.C.: Duke University Press, 2006.

Nading, Alex M. "Local Biologies, Leaky Things, and the Chemical Infrastructure of Global Health." *Medical Anthropology* 36, no. 2 (2017): 141–156.

National Association of Convenience Stores. "Key Facts about Fueling." March 25, 2021. https://www.convenience.org/Topics/Fuels/The-US-Petroleum-Industry-Statistics-Definitions.

National Research Council. *Human Biomonitoring for Environmental Chemicals*. Washington, D.C.: National Academies Press, 2006. https://doi.org/10.17226/11700.

———. *Monitoring Human Tissues for Toxic Substances*. Washington, D.C.: National Academies Press, 1991. https://doi.org/10.17226/1787.

———. *Waste Incineration and Public Health*. Washington, D.C.: National Academies Press, 2000. https://doi.org/10.17226/5803.

Ndiaye, Pap A. *Nylon and Bombs: DuPont and the March of Modern America*. Translated by Elborg Forster. Baltimore: Johns Hopkins University Press, 2007.

Neslen, Arthur. "EU Report on Weedkiller Safety Copied Text from Monsanto Study." *Guardian*, September 14, 2017. https://www.theguardian.com/environment/2017/sep/15/eu-report-on-weedkiller-safety-copied-text-from-monsanto-study.

New York City Department of Parks and Recreation. "Freshkills Park." New York City Department of Parks and Recreation. Accessed June 27, 2020. https://www.nycgovparks.org/park-features/freshkills-park.

Newman, Richard S. *Love Canal: A Toxic History from Colonial Times to the Present*. Oxford: Oxford University Press, 2016.

Nisbet, I. C., and P. K. LaGoy. "Toxic Equivalency Factors (TEFs) for Polycyclic Aromatic Hydrocarbons (PAHs)." *Regulatory Toxicology and Pharmacology* 16, no. 3 (1992): 290–300.

Nixon, Rob. *Slow Violence and the Environmentalism of the Poor*. Cambridge, Mass.: Harvard University Press, 2011.

Noble, Charles. *Liberalism at Work: The Rise and Fall of OSHA*. Philadelphia: Temple University Press, 1986.

Nora, Pierre. *Les lieux de mémoire*. 3 vols. Bibliothèque illustrée des histoires. Paris: Gallimard, 1984.

———. *Realms of Memory: Rethinking the French Past*. Edited by Lawrence D. Kritzman. Translated by Arthur Goldhammer. 3 vols. New York: Columbia University Press, 1996.

Oertel, Angelika, Katrin Maul, Jakob Menz, Anna Lena Kronsbein, Dana Sittner, Andrea Springer, Anne-Katrin Müller, Uta Herbst, Kerstin Schlegel, and Agnes Schulte. *REACH Compliance: Data Availability in REACH Registrations Part 2: Evaluation of Data Waiving and Adaptations for Chemicals ≥ 1000 tpa*. Dessau-Rosslau, Germany: Umweltbundesamt, 2018.

Office of Technology Assessment. *The Information Content of Premanufacture Notices: Background Paper*. Washington, D.C.: U.S. Government Printing Office, 1983.

O'Reilly, James T. "Torture by TSCA: Retrospectives of a Failed Statute." *Natural Resources & Environment* 25, no. 1 (2010): 43–44, 47.

Oreskes, Naomi, and Erik M. Conway. *Merchants of Doubt: How a Handful of Scientists Obscured the Truth on Issues from Tobacco Smoke to Global Warming.* New York: Bloomsbury, 2010.

Organisation for Economic Co-operation and Development. *Extended Producer Responsibility: A Guidance Manual for Governments.* Paris: OECD, 2001.

———. *Trade Measures in Multilateral Environmental Agreements.* Paris: OECD, 1999.

Ottinger, Gwen. *Refining Expertise: How Responsible Engineers Subvert Environmental Justice Challenges.* New York: New York University Press, 2013.

Pagano, Giovanni, Marco Guida, Franca Tommasi, and Rahime Oral. "Health Effects and Toxicity Mechanisms of Rare Earth Elements–Knowledge Gaps and Research Prospects." *Ecotoxicology and Environmental Safety,* 115 (2015): 40–48.

Pallemaerts, Marc. *Toxics and Transnational Law: International and European Regulation of Toxic Substances as Legal Symbolism.* Oxford: Hart, 2003.

Patel, Raj, and Jason W. Moore. *A History of the World in Seven Cheap Things: A Guide to Capitalism, Nature, and the Future of the Planet.* Oakland: University of California Press, 2017.

Patterson, Clair C. "Contaminated and Natural Lead Environments of Man." *Archives of Environmental Health* 11, no. 3 (1965): 344–360.

Paull, Jeffrey M. "The Origin and Basis of Threshold Limit Values." *American Journal of Industrial Medicine* 5, no. 3 (1984): 227–238.

Paustenbach, Dennis, and David Galbraith. "Biomonitoring and Biomarkers: Exposure Assessment Will Never Be the Same." *Environmental Health Perspectives* 114, no. 8 (2006): 1143–1149.

Pearce, Neil, and Ben Caplin. "Let's Take the Heat out of the CKDu Debate: More Evidence Is Needed." *Occupational and Environmental Medicine* 76, no. 6 (2019): 357–359. https://doi.org/10.1136/oemed-2018-105427.

Pellow, David Naguib. *Garbage Wars: The Struggle for Environmental Justice in Chicago* Cambridge, Mass.: MIT Press, 2002.

———. *Resisting Global Toxics: Transnational Movements for Environmental Justice.* Cambridge, Mass.: MIT Press, 2007.

Pestre, Dominique, ed. *Le gouvernement des technosciences. Gouverner le progrès et ses dégâts depuis 1945.* Paris: La Découverte, 2014.

Peto, Julian, John T. Hodgson, Fiona E. Matthews, and Jacqueline R. Jones. "Continuing Increase in Mesothelioma Mortality in Britain." *Lancet* 345, no. 8949 (1995): 535–539.

Pierson, Paul. *Politics in Time: History, Institutions, and Social Analysis.* Princeton, N.J.: Princeton University Press, 2004.

Pilling, Elizabeth. Interview by Lee Sullivan Berry. REACH Ambler, Chemical Heritage Foundation Oral History Collection, Science History Institute, January 6, 2014. Transcript #0825.

Piovano, Pablo Ernesto. "Project: The Human Cost of Agrochemicals." *Manuel Rivera-Ortiz Foundation for Documentary Photography & Film* (blog), April 12, 2018. https://mrofoundation.org/Program-2016.

Pirkle, James L., Debra J. Brody, Elaine W. Gunter, Rachel A. Kramer, Daniel C. Paschal, Katherine M. Flegal, and Thomas D. Matte. "The Decline in Blood Levels in the United States: The National Health and Nutrition Examination Surveys (NHANES)." *Journal of the American Medical Association* 272, no. 4 (1994): 284–291.

Potts, Simon G., Peter Neumann, Bernard Vaissière, and Nicolas J. Vereecken. "Robotic Bees for Crop Pollination: Why Drones Cannot Replace Biodiversity." *Science of the Total Environment* 642 (2018): 665–667.

PR Newswire. "China Rare Earth Industry Report 2017–2021—Research and Markets." June 1, 2017. https://www.prnewswire.com/news-releases/china-rare-earth -industry-report-2017-2021---research-and-markets-300467141.html.

Pritchard, Sara B. "An Envirotechnical Disaster: Nature, Technology, and Politics at Fukushima." *Environmental History* 17, no. 2 (2012): 219–243.

Proctor, Robert N. *Cancer Wars: How Politics Shapes What We Know and Don't Know about Cancer.* New York: Basic Books, 1995.

———. *Golden Holocaust: Origins of the Cigarette Catastrophe and the Case for Abolition.* Berkeley: University of California Press, 2011.

Proctor, Robert N., and Londa L. Schiebinger, eds. *Agnotology: The Making and Unmaking of Ignorance.* Stanford, Calif.: Stanford University Press, 2008.

Puig de la Bellacasa, María. *Matters of Care: Speculative Ethics in More Than Human Worlds.* Minneapolis: University of Minnesota Press, 2017.

Rajapakse, Senaka, Mitrakrishnan Chrishan Shivanthan, and Mathu Selvarajah. "Chronic Kidney Disease of Unknown Etiology in Sri Lanka." *International Journal of Occupational and Environmental Health* 22, no. 3 (2016): 259–264.

Ranasinghe, Sattamabiralalage Maxwell. "Chronic Kidney Disease Unidentified (CKDu) in Sri-Lanka: Towards an Integrated Solution." MA thesis, York University, 2017.

Reinhardt, Carsten. "Limit Values and the Boundaries of Science and Technology." *Comptes Rendus Chimie* 15, no. 7 (2012): 595–602.

———. "Regulierungswissen und Regulierungskonzepte." *Berichte zur Wissenschaftsgeschichte* 33, no. 4 (2010): 351–364.

Reisman, D., R. Weber, J. McKernan, and C. Northeim. *Rare Earth Elements: A Review of Production, Processing, Recycling, and Associated Environmental Issues.* Washington, D.C.: U.S. EPA Office of Research and Development, 2013.

Richards, Linda M. "Rocks and Reactors: An Atomic Interpretation of Human Rights, 1941–1979." PhD diss., Oregon State University, 2014. https://ir.library .oregonstate.edu/concern/graduate_thesis_or_dissertations/vii8rh21s.

Richter, Lauren, Alissa Cordner, and Phil Brown. "Non-stick Science: Sixty Years of Research and (In)action on Fluorinated Compounds." *Social Studies of Science* 48, no. 5 (2018): 691–714.

———. "Producing Ignorance through Regulatory Structure: The Case of Per- and Polyfluoroalkyl Substances (PFAS)." *Sociological Perspectives* 64, no. 4 (2021): 631–656. https://doi.org/10.1177/0731121420964827.

Rijpma, Anouk. "Terres rares: Le pétrole de la Chine." *L'Express*, May 2, 2012. https:// www.lexpress.fr/actualite/societe/terres-rares-le-petrole-de-la-chine_1110422.html.

Roach, S. A., and S. M. Rappaport. "But They Are Not Thresholds: A Critical Analysis of the Documentation of Threshold Limit Values." *American Journal of Industrial Medicine* 17, no. 6 (1990): 727–753.

Robert, Cécile. "Expert Groups in the Building of European Public Policy." *Globalisation, Societies and Education* 10, no. 4 (2012): 425–438.

Roberts, Jody A. "Unruly Technologies and Fractured Oversight: Toward a Model for Chemical Control for the Twenty-First Century." In *Powerless Science? Science and Politics in a Toxic World*, edited by Soraya Boudia and Nathalie Jas, 254–268. New York: Berghahn, 2014.

Robinson, Claire. "The Glyphosate Toxicity Studies You're Not Allowed to See." GM Watch, July 2, 2014. https://gmwatch.org/en/news/archive/2014/15519-the -glyphosate-toxicity-studies-you-re-not-allowed-to-see.

Rosen, George. *A History of Public Health.* New York: MD, 1958.

Rosental, Paul-André. "Before Asbestos, Silicosis. Death from Occupational Disease in Twentieth Century France." *Population & Societies, Publication of the Institut National d'études Démographiques* 437 (September 2007): 1–4.

Roskill Information Services. *Rare Earths: Market Outlook to 2020.* 15th ed. London: Roskill, 2015.

Ross, Benjamin, and Steven Amter. *The Polluters: The Making of Our Chemically Altered Environment.* Oxford: Oxford University Press, 2010.

Rothschild, Rachel Emma. "Burning Rain: The Long-Range Transboundary Air Pollution Project." In *Toxic Airs: Body, Place, Planet in Historical Perspective*, edited by James Rodger Fleming and Ann Johnson, 181–207. Pittsburgh, Pa.: University of Pittsburgh Press, 2014.

———. *Poisonous Skies: Acid Rain and the Globalization of Pollution.* Chicago: University of Chicago Press, 2019.

Rummel, Andreas. *Tote Tiere, kranke Menschen.* Documentary. MDR / Arte, Rumara Fernsehproduktion, 2015.

Schafer, Kristin S., Margaret Reeves, Skip Spitzer, and Susan E. Kegley. *Chemical Trespass: Pesticides in Our Bodies and Corporate Accountability.* Berkeley, Calif.: Pesticide Action Network of North America, 2004.

Schierow, Linda-Jo. *The Toxic Substances Control Act (TSCA): A Summary of the Act and Its Major Requirements.* Washington, D.C.: Congressional Research Service, 2009.

Schwerin, Alexander von. "Low Dose Intoxication and a Crisis of Regulatory Models. Chemical Mutagens in the Deutsche Forschungsgemeinschaft (DFG), 1963–1973." *Berichte zur Wissenschaftsgeschichte* 33, no. 4 (2010): 401–418.

———. "Vom Gift im Essen zu chronischen Umweltgefahren: Lebensmittelzusatzstoffe und die risikopolitische Institutionalisierung der Toxikogenetik in der Bundesrepublik, 1955–1964." *Technikgeschichte* 81, no. 3 (2014): 251–274.

Scruggs, Caroline E., and Leonard Ortolano. "Creating Safer Consumer Products: The Information Challenges Companies Face." *Environmental Science & Policy* 14, no. 6 (October 1, 2011): 605–614. https://doi.org/10.1016/j.envsci.2011.05.010.

Seiler, Claudia, and Thomas U. Berendonk. "Heavy Metal Driven Co-selection of Antibiotic Resistance in Soil and Water Bodies Impacted by Agriculture and Aquaculture." *Frontiers in Microbiology* 3 (2012): 399.

Selin, Henrik. "Coalition Politics and Chemicals Management in a Regulatory Ambitious Europe." *Global Environmental Politics* 7, no. 3 (2007): 63–93.

———. *Global Governance of Hazardous Chemicals: Challenges of Multilevel Management.* Cambridge, Mass.: MIT Press, 2010.

Selin, Henrik, and Nicole Eckley Selin. *Mercury Stories: Understanding Sustainability through a Volatile Element.* Cambridge, Mass.: MIT Press, 2020.

Selin, Noelle Eckley. "Drawing Lessons about Science-Policy Institutions: Persistent Organic Pollutants (POPs) under the LRTAP Convention." Discussion Paper, E-99-11, Belfer Center for Science and International Affairs, Harvard Kennedy School, Cambridge, Mass., June 30, 1999. https://www.belfercenter.org/publication/drawing-lessons-about-science-policy-institutions-persistent-organic -pollutants-pops.

Sellers, Christopher C. *Hazards of the Job: From Industrial Disease to Environmental Health Science.* Chapel Hill: University of North Carolina Press, 1997.

Senier, Laura, Phil Brown, Sara Shostak, and Bridget Hanna. "The Socio-exposome: Advancing Exposure Science and Environmental Justice in a Postgenomic Era." *Environmental Sociology* 3, no. 2 (2017): 107–121. https://doi.org/10.1080/23251042.2016.1220848.

Serres, Michel. *Malfeasance: Appropriation through Pollution?* Translated by Anne-Marie Feenberg-Dibon. Stanford, Calif.: Stanford University Press, 2011.

Sexton, Ken, Larry L. Needham, and James L. Pirkle. "Human Biomonitoring of Environmental Chemicals: Measuring Chemicals in Human Tissues Is the 'Gold Standard' for Assessing People's Exposure to Pollution." *American Scientist* 92, no. 1 (2004): 38–45.

Shapiro, Nicholas, Nasser Zakariya, and Jody Roberts. "A Wary Alliance: From Enumerating the Environment to Inviting Apprehension." *Engaging Science, Technology, and Society* 3 (2017): 575–602.

Sharma, Pratibha. "Delhi's Air: Why Does No One Care about Unmanaged Waste?" *Economic and Political Weekly* 52, no. 50 (December 16, 2017).

Shostak, Sara. *Exposed Science: Genes, the Environment, and the Politics of Population Health.* Berkeley: University of California Press, 2013.

Shotwell, Alexis. *Against Purity: Living Ethically in Compromised Times.* Minneapolis: University of Minnesota Press, 2016.

Silbergeld, Ellen K., Daniele Mandrioli, and Carl F. Cranor. "Regulating Chemicals: Law, Science, and the Unbearable Burdens of Regulation." *Annual Review of Public Health* 36, no. 1 (2015): 175–191.

Silent Spring Institute. "Detox Me Action Kit." Accessed July 3, 2020. https://silentspring.org/detoxmeactionkit.

Sinkkonen, Seija, and Jaakko Paasivirta. "Degradation Half-Life Times of PCDDs, PCDFs and PCBs for Environmental Fate Modeling." *Chemosphere* 40, no. 9 (2000): 943–949.

Sirinathsinghji, Eva, and Mae-Wan Ho. *Why Glyphosate Should Be Banned: A Review of Its Hazards to Health and the Environment.* Nottingham, U.K.: Institute of Science in Society, 2012. http://www.i-sis.org.uk/Why_Glyphosate_Should_be_Banned.php.

Sismondo, Sergio. *Ghost-Managed Medicine: Big Pharma's Invisible Hands.* Manchester: Mattering, 2018.

Slota, Stephen C., and Geoffrey C. Bowker. "How Infrastructures Matter." In *The Handbook of Science and Technology Studies,* edited by Ulrike Felt, Rayvon Fouché, Clark A. Miller, and Laurel Smith-Doerr, 529–554. 4th ed. Cambridge, Mass.: MIT Press, 2016.

Spears, Ellen Griffith. *Baptized in PCBs: Race, Pollution, and Justice in an All-American Town.* Chapel Hill: University of North Carolina Press, 2014.

Stadler, Linda. "Corrosion Proof Fittings v. EPA: Asbestos in the Fifth Circuit—a Battle of Unreasonableness." *Tulane Environmental Law Journal* 6, no. 2 (1993): 423–438.

Stayner, Leslie, Laura S. Welch, and Richard Lemen. "The Worldwide Pandemic of Asbestos-Related Diseases." *Annual Review of Public Health* 34 (2013): 205–216.

Steingraber, Sandra. *Living Downstream: A Scientist's Personal Investigation of Cancer and the Environment.* New York: Vintage, 1998.

Stokstad, Erik. "Pollution Gets Personal." *Science* 304, no. 5679 (2004): 1892–1894. https://doi.org/10.1126/science.304.5679.1892.

Strasser, Susan. *Waste and Want: A Social History of Trash.* New York: Metropolitan, 1999.

Summit Realty Advisors, LLC. "Ambler Boiler House." Accessed June 27, 2020. https://www.summitrealtyadvisors.com/ambler-boiler-house.

Suryanarayanan, Sainath, and Daniel Lee Kleinman. "Be(e)coming Experts: The Controversy over Insecticides in the Honey Bee Colony Collapse Disorder." *Social Studies of Science* 43, no. 2 (2013): 215–240. https://doi.org/10.1177/0306312712466186.

———. *Vanishing Bees: Science, Politics, and Honeybee Health.* New Brunswick, N.J.: Rutgers University Press, 2016.

Szasz, Andrew. *Shopping Our Way to Safety: How We Changed from Protecting the Environment to Protecting Ourselves.* Minneapolis: University of Minnesota Press, 2007.

Sze, Julie. *Noxious New York: The Racial Politics of Urban Health and Environmental Justice.* Cambridge, Mass.: MIT Press, 2006.

Tainter, Joseph A. "Global Change, History, and Sustainability." In *The Way the Wind Blows: Climate, History, and Human Action,* edited by Roderick J. McIntosh, Joseph A. Tainter, and Susan Keech McIntosh, 331–356. New York: Columbia University Press, 2000.

Tarr, Joel A. *The Search for the Ultimate Sink: Urban Pollution in Historical Perspective.* Akron, Ohio: University of Akron Press, 1996.

Tauveron, Albert. *Les années poubelle.* Grenoble, France: Presses Universitaires de Grenoble, 1984.

Thébaud-Mony, Annie. "Les fibres courtes d'amiante sont-elles toxiques? Production de connaissances scientifiques et maladies professionnelles." *Sciences sociales et sante* 28, no. 2 (2010): 95–114.

Thornton, Joseph W., Michael McCally, and Jane Houlihan. "Biomonitoring of Industrial Pollutants: Health and Policy Implications of the Chemical Body Burden." *Public Health Reports* 117, no. 4 (2002): 315–323.

Thorsheim, Peter. *Inventing Pollution: Coal, Smoke, and Culture in Britain since 1800.* Athens: Ohio University Press, 2006.

"Toxic Substances Control Act of 1976." *Code of Federal Regulations,* 15 U.S.C., §§2601–2629 (1976).

Travis, Anthony S. *The Rainbow Makers: The Origins of the Synthetic Dyestuffs Industry in Western Europe.* Bethlehem, Pa.: Lehigh University Press, 1993.

Trischler, Helmuth. "The Anthropocene: A Challenge for the History of Science, Technology, and the Environment." *NTM Zeitschrift für Geschichte der Wissenschaften, Technik und Medizin* 24, no. 3 (2016): 309–335.

Tsing, Anna Lowenhaupt. *The Mushroom at the End of the World: On the Possibility of Life in Capitalist Ruins.* Princeton, N.J.: Princeton University Press, 2015.

Tweedale, Geoffrey. *Magic Mineral to Killer Dust: Turner & Newall and the Asbestos Hazard.* Oxford: Oxford University Press, 2000.

Uekötter, Frank, and Uwe Lübken, eds. *Managing the Unknown: Essays on Environmental Ignorance.* New York: Berghahn, 2014.

United Nations Environment Programme. *Global Report on the Status of Legal Limits on Lead in Paint.* Nairobi, Kenya: UNEP, 2016.

———. *Recycling Rates of Metals: A Status Report.* Nairobi, Kenya: UNEP, 2011.

U.S. Bureau of Labor Statistics. "Quarterly Census of Employment and Wages." In *Number of Establishments in Private NAICS 447 Gasoline Stations for All Establishment Sizes in U.S. TOTAL, NSA.* Washington, D.C.: U.S. Department of Labor. Accessed July 28, 2020. https://data.bls.gov/timeseries/ENUUS000205447?amp%253bdata_tool=XGtable&output_view=data&include_graphs=true.

U.S. Congress, House, Committee on Interstate and Foreign Commerce, Subcommittee on Oversight and Investigations. *Waste Disposal Site Survey Report Together with Additional and Separate Views, 96-IFC 33.* 96th Congress, 1st Session. Washington, D.C.: U.S. Government Printing Office, 1979.

U.S. Congress, Senate, Committee on Environment and Public Works, Subcommittee on Superfund and Environmental Oversight to the Committee on Environment and Public Works. *Implementation of the Toxic Substances Control Act, the PCB Rule, and Federal Hazardous Substance Laws, Concerning the Performance of the Environmental Protection Agency in the Matter of the Texas Eastern Gas Pipeline Company.* Washington, D.C.: U.S. Government Printing Office, 1988.

U.S. Department of Energy. *Critical Materials Strategy.* Washington, D.C.: U.S. DOE, 2010. http://www.osti.gov/servlets/purl/1219038/.

———. *Critical Materials Strategy.* Washington, D.C.: U.S. DOE, 2011.

U.S. Environmental Protection Agency. "A Citizen's Guide to Monitored Natural Attenuation." Overviews and Factsheets. U.S. EPA, September 2012. https://www.epa.gov/remedytech/citizens-guide-monitored-natural-attenuation.

———. "ACE: Biomonitoring—Polychlorinated Biphenyls (PCBs)." U.S. EPA, May 19, 2015. https://www.epa.gov/americaschildrenenvironment/ace-biomonitoring-polychlorinated-biphenyls-pcbs.

———. "EPA Bans PCB Manufacture; Phases Out Uses." U.S. EPA, April 19, 1979. https://archive.epa.gov/epa/aboutepa/epa-bans-pcb-manufacture-phases-out-uses.html.

———. "Leaking Underground Storage Tanks Corrective Action Resources." Overviews and Factsheets. U.S. EPA, December 8, 2014. https://www.epa.gov/ust/leaking-underground-storage-tanks-corrective-action-resources.

———. *Unfinished Business: A Comparative Assessment of Environmental Problems, Overview Report.* Washington, D.C.: Office of Policy Planning and Evaluation, EPA, 1987.

U.S. Geological Survey. "Asbestos Statistics and Information." National Minerals Information Center. Accessed July 3, 2020. https://www.usgs.gov/centers/nmic/asbestos-statistics-and-information.

———. *Mineral Commodity Summaries.* Washington, D.C.: USGS, 2013.

———. *Mineral Commodity Summaries.* Washington, D.C.: USGS, 2020.

U.S. General Accounting Office. *Toxic Chemicals: Long-Term Coordinated Strategy Needed to Measure Exposures in Humans.* HEHS-00-80. Washington, D.C.: U.S. GAO, 2000.

U.S. Right to Know. "Roundup (Glyphosate) Cancer Cases: Key Documents & Analysis." Monsanto Papers. Accessed July 3, 2020. https://usrtk.org/monsanto-papers/.

van Sittert, N. J., and G. de Jong. "Biomonitoring of Exposure to Potential Mutagens and Carcinogens in Industrial Populations." *Food and Chemical Toxicology* 23, no. 1 (1985): 23–31.

Vogel, Sarah A. *Is It Safe? BPA and the Struggle to Define the Safety of Chemicals.* Berkeley: University of California Press, 2013.

Vogel, Sarah A., and Jody A. Roberts. "Why the Toxic Substances Control Act Needs an Overhaul, and How to Strengthen Oversight of Chemicals in the Interim." *Health Affairs* 30, no. 5 (May 2011): 898–905.

Wajcman, Judy. *Pressed for Time: The Acceleration of Life in Digital Capitalism.* Chicago: University of Chicago Press, 2015.

Walker, Brittania. *Killing Them Softly . . . Health Effects in Arctic Wildlife Linked to Chemical Exposures.* Edited by Julian Woolford and Noemi Cano. Oslo, Norway: WWF International Arctic Programme & WWF-DetoX, 2006.

Walker, J. Samuel. *Permissible Dose: A History of Radiation Protection in the Twentieth Century.* Berkeley: University of California Press, 2000.

Wang, Zhanyun, Glen W. Walker, Derek C. G. Muir, and Kakuko Nagatani-Yoshida. "Toward a Global Understanding of Chemical Pollution: A First Comprehensive Analysis of National and Regional Chemical Inventories." *Environmental Science & Technology* 54, no. 5 (2020): 2575–2584.

Warren, Christian. *Brush with Death: A Social History of Lead Poisoning.* Baltimore: Johns Hopkins University Press, 2000.

Washburn, Rachel. "The Social Significance of Human Biomonitoring." *Sociology Compass* 7, no. 2 (2013): 162–179.

Weeks, Ruth. Interview by Lee Sullivan Berry. REACH Ambler, Chemical Heritage Foundation Oral History Collection, Science History Institute, August 14, 2014. Transcript #0827.

Wijedasa, Namini. "It's Official: Glyphosate Import Is Banned." *Sunday Times Sri Lanka,* June 14, 2015. http://www.sundaytimes.lk/150614/news/its-official -glyphosate-import-is-banned-153388.html.

Williams, E. Spencer, Julie Panko, and Dennis J. Paustenbach. "The European Union's REACH Regulation: A Review of Its History and Requirements." *Critical Reviews in Toxicology* 39, no. 7 (2009): 553–575.

World Wildlife Fund. "Bad Blood? WWF Reveals that Ministers Are Contaminated with an Average of at Least 37 Chemicals." October 19, 2004. https://www.wwf.eu/ ?15893/.

———. "European Parliamentarians Contaminated by 76 Chemicals." April 21, 2004. https://wwf.panda.org/wwf_news/?12622/European-parliamentarians -contaminated-by-76-chemicals.

Wylie, Sara Ann. *Fractivism: Corporate Bodies and Chemical Bonds.* Durham, N.C.: Duke University Press, 2018.

Wylie, Sara, Nick Shapiro, and Max Liboiron. "Making and Doing Politics through Grassroots Scientific Research on the Energy and Petrochemical Industries." *Engaging Science, Technology, and Society* 3 (2017): 393–425.

Yant, W. P., H. H. Schrenk, and F. A. Patty. "A Plant Study of Urine Sulfate Determinations as a Measure of Benzene Exposure." *Journal of Industrial Hygiene and Toxicology* 18 (1936): 349–356.

Zaharchuk, John. Interview by Lee Sullivan Berry. REACH Ambler, Chemical Heritage Foundation Oral History Collection, Science History Institute, January 22, 2014. Transcript #0828.

Zalasiewicz, Jan, Mark Williams, Will Steffen, and Paul Crutzen. "The New World of the Anthropocene." *Environmental Science & Technology* 44, no. 7 (2010): 2228–2231.

Zerubavel, Eviatar. *Time Maps: Collective Memory and the Social Shape of the Past.* Chicago: University of Chicago Press, 2003.

Zierler, David. *The Invention of Ecocide: Agent Orange, Vietnam, and the Scientists Who Changed the Way We Think about the Environment.* Athens: University of Georgia Press, 2011.

Index

Page numbers followed by *f* refer to figures.

About the Authors

SORAYA BOUDIA is a professor of sociology at the University of Paris, specializing in science, environment, and society studies. Her last book, with Nathalie Jas, *Gouverner un monde toxique* analyzes the different ways of governing our toxic world. She has edited four books exploring the relationships between science, expertise, and politics in various contexts, most recently with Nathalie Jas, *Powerless Science? Science and Politics in a Toxic World,* and with Emmanuel Henry, *La Mondialisation des risques: Une histoire politique et transnationale des risques sanitaires environnementaux.* Her current work focuses on the politics of resources and circular economy in the age of global environmental crisis.

ANGELA N. H. CREAGER is the Thomas M. Siebel Professor in the History of Science at Princeton University, where she specializes in the history of biology and biomedicine. She is the author of two books, both published by the University of Chicago Press, most recently *Life Atomic: A History of Radioisotopes in Science and Medicine.* She has also edited four volumes on various topics including, with Jean-Paul Gaudillière, *Risk on the Table: Food Production, Health, and the Environment.* Her current work focuses on the role of genetic tests in environmental science and regulation during the late twentieth century.

SCOTT FRICKEL is a professor of sociology and environment and society at Brown University. He is the author of *Chemical Consequences: Environmental Mutagens and the Rise of Genetic Toxicology*, published by Rutgers University Press, and with James R. Elliott, a comparative history of urban industrial land use entitled *Sites Unseen: Uncovering Hidden Hazards in American Cities*. He has also coedited three books exploring the politics of science, fields of knowledge, and interdisciplinary collaboration. He is currently studying the relationship between hazardous land uses, regulatory science, inequality, and health in Argentina and the United States.

EMMANUEL HENRY is a professor of sociology at University Paris-Dauphine, PSL University, and is a 2020–2021 member of the School of Social Science of the Institute for Advanced Study, Princeton, N.J. He is the author of two books, most recently *Ignorance scientifique et inaction publique: Les politiques de santé au travail*. He has also edited or coedited six books on the construction of social problems, scientific expertise, environmental risks and the power of corporations in public policies. He recently coedited the special issue of *Science, Technology, & Human Values*: "Beyond the Production of Ignorance: The Pervasiveness of Industry Influence through the Tools of Chemical Regulation" (with Valentin Thomas, Sara Aguiton, Marc-Olivier Déplaude and Nathalie Jas). He is currently working on the links between scientific knowledge (and ignorance), expertise, and public policy, mostly in the fields of environmental and occupational health in France and in Europe.

NATHALIE JAS is a senior researcher at the French National Research Institute for Agriculture, Food and Environment (INRAE). As an STS scholar, she has been working extensively on issues related to toxicants, including pesticides. She is the author of *Au Carrefour de la chimie et de l'agriculture: Les sciences agronomiques en Allemagne et en France 1840–1914*, and with Soraya Boudia, a sociohistory of the government of our toxic world entitled *Gouverner un monde toxique*.

She has also coedited three special issues and four books, including *Toxicants, Health and Regulation since 1945* (with Soraya Boudia).

CARSTEN REINHARDT is a professor for historical studies of science at the University of Bielefeld, where he is a director at both the Institute for Interdisciplinary Studies of Science (I²SoS) and the Center for Interdisciplinary Research (ZiF). Between 2013 and 2016, he was president and CEO of the Science History Institute in Philadelphia, USA. Reinhardt's research fields are the history of chemistry, industrial research, and scientific instrumentation and regulation. He authored and coauthored three books, among them *Shifting and Rearranging: Physical Methods and the Transformation of Modern Chemistry*. Most recently, he coedited with Ursula Klein the volume *Objects of Chemical Inquiry*, and with Marianne Sommer and Staffan Müller-Wille, the *Handbuch Wissenschaftsgeschichte*. His current work focuses on expert knowledge and regulation in its interaction with politics and the history of the Max Planck Society.

JODY A. ROBERTS is an independent scholar in Philadelphia. Roberts's work has experimented with ways in which we bring the intellectual core of science studies into the ideas, expectations, and experiences of everyday life. His current work explores the intersection of innovation, imagination, and disability. Previously, he served as the director of the Institute for Research at the Science History Institute, an innovation lab for historical and social science methods designed to strengthen stakeholder engagements in processes of imagining and building new futures. His written and multimedia work include "From Inception to Reform: An Oral History of the Toxic Substances Control Act" with Christy Schneider and Elizabeth McDonnell, the exhibition "Sensing Change," the "REACH Ambler" project, and "A Wary Alliance: From Enumerating the Environment to Inviting Apprehension" with Nicholas Shapiro and Nasser Zakariya in *Engaging STS*.

Available titles in the Nature, Society, and Culture series

Diane C. Bates, *Superstorm Sandy: The Inevitable Destruction and Reconstruction of the Jersey Shore*

Soraya Boudia, Angela N. H. Creager, Scott Frickel, Emmanuel Henry, Nathalie Jas, Carsten Reinhardt, and Jody A. Roberts, *Residues: Thinking through Chemical Environments*

Elizabeth Cherry, *For the Birds: Protecting Wildlife through the Naturalist Gaze*

Cody Ferguson, *This Is Our Land: Grassroots Environmentalism in the Late Twentieth Century*

Shaun A. Golding, *Electric Mountains: Climate, Power, and Justice in an Energy Transition*

Aya H. Kimura and Abby Kinchy, *Science by the People: Participation, Power, and the Politics of Environmental Knowledge*

Anthony B. Ladd, ed., *Fractured Communities: Risk, Impacts, and Protest against Hydraulic Fracking in U.S. Shale Regions*

Stefano B. Longo, Rebecca Clausen, and Brett Clark, *The Tragedy of the Commodity: Oceans, Fisheries, and Aquaculture*

Stephanie A. Malin, *The Price of Nuclear Power: Uranium Communities and Environmental Justice*

Kari Marie Norgaard, *Salmon and Acorns Feed Our People: Colonialism, Nature, and Social Action*

J. P. Sapinski, Holly Jean Buck, and Andreas Malm, eds., *Has It Come to This? The Promises and Perils of Geoengineering on the Brink*

Chelsea Schelly, *Dwelling in Resistance: Living with Alternative Technologies in America*

Sara Shostak, *Back to the Roots: Memory, Inequality, and Urban Agriculture*

Diane Sicotte, *From Workshop to Waste Magnet: Environmental Inequality in the Philadelphia Region*

Sainath Suryanarayanan and Daniel Lee Kleinman, *Vanishing Bees: Science, Politics, and Honeybee Health*

Patricia Widener, *Toxic and Intoxicating Oil: Discovery, Resistance, and Justice in Aotearoa New Zealand*

Printed in the United States
by Baker & Taylor Publisher Services